Exploring Environmental Science with GIS

An Introduction to Environmental Mapping and Analysis

Meg E. Stewart
Mary Ann Cunningham
Jill S. Schneiderman
Liv Gold

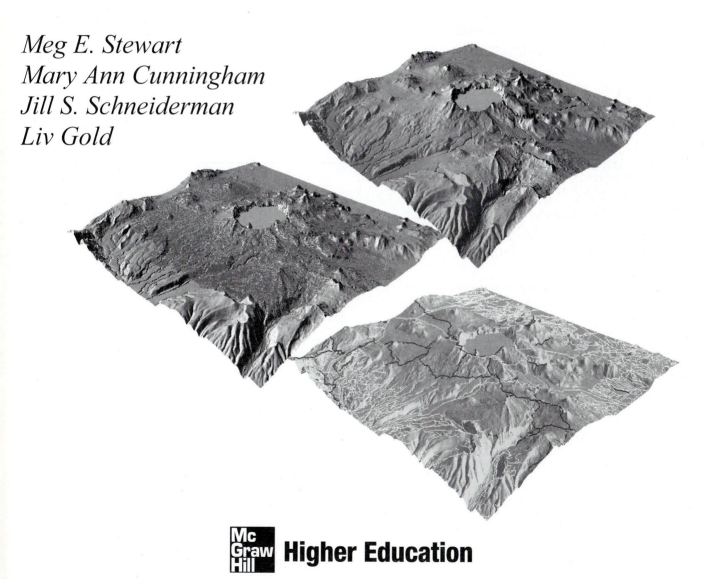

Mc Graw Hill **Higher Education**

Boston   Burr Ridge, IL   Dubuque, IA   Madison, WI   New York   San Francisco   St. Louis
Bangkok   Bogotá   Caracas   Kuala Lumpur   Lisbon   London   Madrid   Mexico City
Milan   Montreal   New Delhi   Santiago   Seoul   Singapore   Sydney   Taipei   Toronto

Higher Education

EXPLORING ENVIRONMENTAL SCIENCE WITH GIS: AN INTRODUCTION TO ENVIRONMENTAL MAPPING AND ANALYSIS

1 2 3 4 5 6 7 8 9 0 QPD/QPD 0 9 8 7 6 5 4

ISBN 0–07–297564–4

Publisher: *Margaret J. Kemp*
Senior developmental editor: *Kathleen R. Loewenberg*
Executive marketing manager: *Lisa L. Gottschalk*
Lead project manager: *Joyce M. Berendes*
Senior production supervisor: *Sherry L. Kane*
Senior media technology producer: *Jeffry Schmitt*
Cover designer: *Laurie B. Janssen*
(USE) Cover image: *Mary Ann Cunningham*
Compositor: *Interactive Composition Corporation*
Typeface: *11/12 Times Roman*
Printer: *Quebecor World Dubuque, IA*

www.mhhe.com

Table of Contents

Preface

Welcome to *Exploring Environmental Science with GIS*. Our aim in this book is to provide an introduction to environmental investigation and problem solving using Geographic Information Systems (GIS). You will investigate environmental and geographic data, and you will practice basic techniques of spatial analysis in environmental science.

What Is GIS?

A GIS is spatially explicit data—points, lines, or polygons with known geographic coordinates—and software used to display, overlay, modify, store, and manipulate those data. Points, lines, and polygons are defined to represent features in the landscape, such as data collection points, roads, or zoning areas in a county. Each point, line, or polygon has attached descriptive information. A data collection point, for example, might be a weather station. Data such as rainfall rates, temperatures, wind speeds, and barometric pressure are recorded at that point, and a table of records can be attached to a set of points. Rainfall patterns across a region can be mapped if a number of data collection points are available.

GIS has become a key tool in many environmental disciplines. Ecologists, planners, geologists, environmental chemists, urban planners, political scientists, and, of course, geographers, all make use of the maps and spatially distributed data used in GIS. This book is intended to provide some experience to people who have no experience with GIS or with geographic data, but who are interested in learning about it. We hope you will learn some of the ways that GIS can be used to identify, explore, understand, and solve problems in environmental science.

It goes almost without saying that environmental phenomena are spatially distributed. Aquifer contamination occurs where porous rock underlies agricultural or industrial land uses. Erosion occurs where soft sediments occur on slopes, especially if those slopes are cleared or developed. Often we understand these relationships in principle, but we understand them more clearly, and more thoroughly, when we see them on a map. Visualization, being able to "see" spatial relationships and patterns in the landscape, can make a great difference in our understanding of environmental problems and solutions. An ability to calculate the spatial extent of overlapping variables, such as soft sediment and land clearance, is key to analyzing environmental problems.

Who Should Use This Book?

This workbook is intended to serve students in a wide range of courses and disciplines. Exercises engage issues at the intersection of many disciplines, such as water quality, population growth, environmental hazards, and land use. Exercises may be done in any order, with the exception of the first one, which introduces basic functions in ArcExplorer. The length of exercises is designed so that most students can complete their work in about an hour, and we have written exercises so that most students will be able to work without instructor supervision. The time spent on each exercise will depend, of course, on students' individual working pace. All exercises ask that students answer questions as they work or print maps to show their end results. Instructors may collect worksheets from this workbook, or final maps, as part of student assignments.

In addition to basic GIS functions, exercises ask that students create graphs from data exported from their maps. Some exercises allow students to open projects directly at the beginning of each chapter while others require students to build projects from the data provided. Students are also asked to export images of their maps, which can be handed in as a conclusion to exercises.

The Software: ArcExplorer

There are many GIS software packages, and many formats of spatial data. The exercises in this book use ArcExplorer, a program that performs many standard GIS operations and that can be downloaded free from Environmental Systems Research, Inc. (ESRI). Full-featured GIS packages have more functions than does ArcExplorer, but this free software provides a no-cost opportunity for new users to learn about what a GIS can do for them, how it works, and what kind of data and mapping can be done with this technology. Because ESRI makes ArcExplorer available free of charge, we ask that students acquire the program directly from the ESRI website. Downloading and installation should take between 15 and 30 minutes, depending on modem speed and individual work speed. After ArcExplorer is installed, data for all exercises can be copied directly from the CD in this book to students' computers.

Students who find these exercises enjoyable or stimulating can explore opportunities for further training in GIS. Courses are available at many colleges and universities. This technology is growing, exciting, and very widely useful.

System Requirements

To complete the exercises in this book, students will need a computer with a CD drive and Internet access (for initial installation of ArcExplorer). The software will use 20.5 mb of hard drive space. If students copy the entire contents of the CD to their hard drives, they will need 450 mb of disk space. Students may prefer to copy data for one exercise at a time. If students have access to a printer, printed maps can be collected along with student answers to the exercises in this workbook.

The exercises in this book are written for use on a Windows personal computer. ArcExplorer for Macintosh is now available, but file management differs between the two systems, so that these exercises function best in Windows. Future editions of this manual will provide exercises for Macintosh computers.

About the Authors

The authors of this workbook are members of the Department of Geology and Geography at Vassar College. All have used GIS in their courses as well as in research activities. Meg Stewart is the Academic Computing Consultant for GIS. Her research interests include environmental justice and environmental health. Mary Ann Cunningham is an Assistant Professor of Geography, with research foci in biogeography and landscape ecology. Jill Schneiderman is a Professor of Geology. Her research interests include water resources, environmental justice, sedimentary geology, and history of science. Liv Gold studies Environmental Studies and Geology at Vassar.

Because we regularly collaborate with colleagues in a wide range of disciplines, we are especially interested in finding ways to share the tools of GIS with students and scholars with a range of backgrounds. We hope that the exercises in this book will show some of the ways the software can be used to investigate a broad variety of questions and problems.

Acknowledgements

We owe thanks to many people who made this project possible. Our families provided extraordinary amounts of emotional and logistical support during this process. Marnie Archer, Craig Dalton, Jake Hoffman, Ryan Lamb, and Krysia Skorko assisted with data acquisition and project development. We are also grateful to Marge Kemp, Managing Editor of Earth and Environmental Science, as well as Kathy Loewenberg, Senior Developmental Editor, and Joyce Berendes, Lead Project Manager, of McGraw-Hill for their work on this project.

Installation: Getting Started with ArcExplorer

ArcExplorer is the GIS software package used in this manual. Environmental Systems Research, Inc. (ESRI) provides this software free of charge. Before you begin this manual, you will need to go to the ESRI website to download the software. Downloading and installation involve several steps, but they should take most users between 15 and 30 minutes, depending on your work speed and your Internet connections. Please note that if you are working in a student lab, you may need administrator privileges to install this software.

1. *Go to the ESRI Products website:* Open a web browser and type http://www.esri.com/arcexplorer.html in the address bar.

2. *Navigate to ArcExplorer:* At the top of the page is a graphic roll-over menu like that shown to the right. Click on "Downloads" to open a page listing the available versions of software. Under "Available Versions," click on the first link: ArcExplorer 4.0.1 - Java Edition: Download with Instructions.

3. *Choose your operating system and download ArcExplorer:* There are several notes at the top of the downloading page. Below these you will find Instructions for downloading. Click on the link the Windows operating system.

   You will see a list of system requirements and the link to Download JRE and ArcExplorer 4.0.1 for Windows. Click on this link to begin downloading.

4. *Register with ESRI before downloading:* Click on the word Register when it appears. By keeping track of where their software is used, and how many copies are in use, ESRI can do a better job of maintaining and updating the software.

   Fill in the appropriate information and click on the "Register Me!" box. To continue, click download.

   Type your email address in the space provided and click enter. This should lead to a page with the message, "(Your first name), thanks for your interest in ESRI software."

   Below this message are instructions for selecting a downloading link with a nearby location. In the box below, choose a location close to you, and click on its corresponding link.

A File Download window will appear. If asked whether to run the program from its current location or save to disk, select "Save this file to disk," (as shown below), then click OK. You will be downloading a compressed (".zip") file that will be used to install the software on your computer. Save this file on your desktop. (You may save this compressed file anywhere, but make sure you know where you save it, so that you can find it once it is downloaded.)

Note that steps in downloading the file may differ somewhat, depending on your version of the Windows operating system and your browser. When the download is complete, close the dialogue box by clicking the "Close" button.

5. *Install ArcExplorer:* Locate the downloaded file and double click on it. This will open a decompression program such as Winzip or PKZIP. If you do not have one of these decompressing programs, you can pause and download it. Please see the note at the end of this section. Extract or decompress the files. Extract them to your desktop.

    After extracting the files, find them on your desktop. Double click on "j2re-1_4_0-win-i.exe." This will install Java, which will make ArcExplorer run more efficiently. As you run this installation, you will be asked to choose a destination location for the Java program. Accept the default location (C:\Program Files\Java\j2re1.4.0).

    Once JRE is installed, you can install ArcExplorer. On your desktop, double click the AEJavaSetup.exe icon. Once again, accept the default location for this program: C:\Program Files\ArcGIS\ArcIMS.

6. *Open a map project in ArcExplorer:* To view a map, you will need to copy data from your CD to your computer. Projects will not run properly directly from the CD. First you must make a folder called ExploringGIS to hold the data: double click on the "My Computer" icon on your desktop. Then double click on the Local Disk (C:). Then click on the File menu, and select New > Folder. Name your new folder "ExploringGIS." (Note that if you use another name, some of the projects may not open correctly.)

    Now insert your CD into your computer's CD drive, then copy the folder **World_Demographics** from your CD to your new ExploringGIS folder.

Once you have copied the **World_Demographics** folder to your hard drive, you can begin to use ArcExplorer. Click on the Start button, then select ArcGIS > ArcExplorer 4.0.1. This will start the software.

In the ArcExplorer toolbar, find the Open Project icon (or go to the File menu > Open Project):

Navigate to the **World_Demographics** folder you just copied to your computer. In that folder you'll find the file **World_Demographics.axl**. The ".axl" extension indicates that this is an ArcExplorer project file. Double click on the name of this file to open the map project.

You have now installed ArcExplorer and opened a map project in the program. You may stop now, or you may continue to Exercise 1: Mapping World Demographics. If you wish to stop now, close ArcExplorer by clicking on the "X" in the upper right corner of the window.

Note Regarding Decompression Software
If you are not sure whether you have Winzip, PKZip, or another decompression program on your computer, the easiest way to check is to double click on a compressed file. You can identify compressed files by their three-letter file name extensions, usually ".zip." Your computer should find the decompression software if you have it.

If your computer asks how to open the file, you can either consult a system administrator or go to Pkware.com to download a shareware version of the software. Go to www.pkware.com, then find Products, then look for Free Downloads. You may also purchase the software, which is relatively inexpensive.

Exercise 1: Mapping World Demographics

One of the most exciting aspects of working with GIS is the ability to make your own maps comparing variables among places. In this exercise, you'll make a map of global demographic factors. Along the way, you'll become familiar with the basic functions of ArcExplorer. Unlike Chapter 2, this exercise calls for you to create your own project. The process by which you do this is detailed in the next paragraph.

Create a Folder and Copy Data from the CD
If you have not already done so, create a new folder on your C:\ drive, and call it **ExploringGIS**. Then copy the **World_Demographics** folder from your CD to the new ExploringGIS folder.

Note: For all exercises in this workbook, you should begin by copying data from the CD to the ExploringGIS folder you just created. Plan to store your data in this folder, and save your results here as you work. As you go through the exercises, remember that to open correctly, the data should always be copied to this C:\ExploringGIS\folder.

In the **World_Demographics** folder, you'll find several files. All share the name World_Demographics, but they have different 3-letter extensions (e.g., World_Demographics.shp, World_Demographics.dbf). Any file you use in ArcExplorer will come as a set of files like this one. Usually, we refer to this group of files as a shape file, because it tells ArcExplorer how to draw the shapes of geographic areas. In this case, it shows countries of the world.

Open ArcExplorer
If you haven't already done so, open ArcExplorer: click on the Start button, then select ArcGis > ArcExplorer 4.0.1 (on a PC), or on a Macintosh, locate the ArcExplorer program using the Finder and double click on its name or icon.

(Note: It is usually best to copy your data before opening ArcExplorer. The program may not find files added after it was opened.)

Add a File
Click on the Layers menu > Add Layers (or click Ctrl-A). You can also click on the red and yellow "+" icon near the top of the ArcExplorer window.

Note: On some computers, the following error message may appear when adding files. If you see this message, just press Continue or Cancel and go on.

Navigate to find the **World_Demographics** files you just copied from the CD.

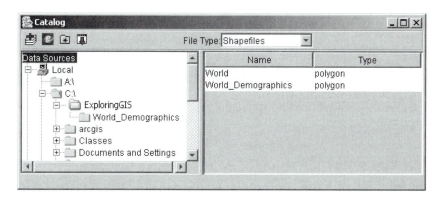

Add the file **World_Demographics**: double click on its name, or press the red and yellow "+" icon. Close the catlog window by clicking on the X in the upper right corner.

Now click on the small box next to the layer's name to turn on the layer.

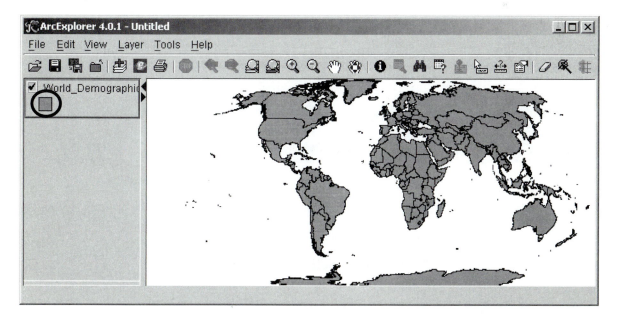

You can now see the countries of the world, but it would be nice to have an outline of a globe for background. Add the **World** layer by repeating the steps you just performed.

When you turn on the **World** layer, you lose the countries. Why? This happens because the **World** layer is on top of the countries layer.

Try changing the order of the layers. Select the **World_Demographics** layer: click on its name, and you will see a dark box around the layer's name and symbol.

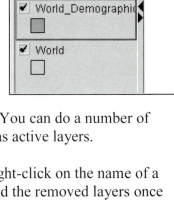

Once you have selected a layer, you can drag it up and down in the legend (the list of layers). You can look at its attributes. You can change its colors. You can do a number of operations on layers once they are selected. Selected layers are also known as active layers.

While you are adding files, you might want to know how to remove one. Right-click on the name of a layer (on a PC), and select Remove Layer. Do this once for practice, then add the removed layers once again to your map.

Note: You can hide the legend or the map by clicking on the small arrows in its margin. Try clicking these arrows so you'll understand how they work. You can also drag this bar left and right.

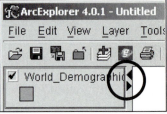

Also note that you may encounter error messages when adding files.

Look at Feature Attributes

Each polygon (country) has a number of statistics that you can map. Let's look at what kind of statistics there are. First select a country: click on the Select Features tool, and select any shape.

Select
Features

With this tool, you can click and drag a rectangle or circle, and all features *in the active layer* that intersect the square or circle will be selected. You can also click on a single polygon. Click on a single country for now.

Try changing your selection: click the Clear All Selection icon:

Clear All
Selection

Now select a country again, and this time click on the Attributes icon.

Attributes

Note that the right-most icons are only available when a feature is selected. Many other icons are only available *when there is an active layer*.

When you click on the Attributes icon, you should see a list of variables, or Fields, and the value of each variable for the country you've selected. In the example below, Canada is selected, and the country's population for a number of years is shown.

Field	Value
COUNTRY	Canada
POP1950	14011422
POP1960	18266765
POP1970	21749986
POP1980	24593300
POP1990	27790600
POP2000	31278097
POP2010	34252514

Attributes — 1 feature — Canada — Layer: World_Demographics

Note that you can stretch the Attributes window to make it longer and see more variables, or Fields, at once. Also, you can scroll up and down using the scroll bar at right.

Look down the list to see what kind of variables you have to work with.
Then close the window.

Let's Map Some Attributes

If you double click on the active layer, or press Ctrl-L, you will see the Symbols window. Initially, layers are displayed using one symbol—that is, one color or line type is used for all features in the layer.

Let's change the color and style. First change the color to Cyan. You can see if you like the new color by clicking the Apply button.

Now try changing the fill style. Notice that by choosing different fills you can see lower layers better. You can also change the outline style and color, or remove the outline altogether, in the lower part of the window.

Reset the Symbol

In most map making, solid fills are easier on the eye. Change the fill back to solid, in a color you like, and change the outlines back to solid, black lines. These are easiest to see.

Now change to Unique Symbols using the top drop-down menu. Unique Symbols assign a unique color to each feature, or each value, in the map. For "Field for values" select CNTRY_NAME.

This method should give a different color to each country. Note that there may be multiple polygons with a single color, as in the case of Indonesia.

Unique Symbols is useful, but Graduated Symbols, or Graduated Colors, are even more helpful for comparing variables between countries.

Change to Graduated Symbols. This mapping method breaks the range of values into classes. You can set the number of classes, but usually 5 or 6 is a good choice. First use POP1950 as the field you are mapping, and set it to 4 or 5 classes. Note that you can select the top and bottom colors, and the program will assign intermediate colors for you.

Now look at your map.
Can you see interesting patterns of high and low values?

Answer:_____

Which country had the highest population in 1950? Answer:_____

The population map is not especially informative. There are two problems with it. One is that it shows *total values,* the number of people per country. Since large countries often have large populations, mapping total values by color classes can be confusing: for example, how do you know whether a population is dense or simply expansive? A good rule to remember is that you should never map total values using this area-shading ("choropleth") mapping method. Instead, map the *rates,* such as population per square kilometer or births per 1,000 people.

A second reason for the lack of information in your population map is that the data range is not well represented. You cannot see a good deal of the distribution because one or two countries have such extremely high values that the rest of the world falls in the bottom one or two classes. Other mapping programs let you choose a classification method: quantiles, natural breaks, or standard deviations, to name a few, but ArcExplorer restricts you to equal intervals. (Considering the cost of the software, perhaps this is only fair.) This means that you should think carefully about the variables that you choose to map.

To correct the total values problem, try mapping DEN1950 (open the Symbols Properties window again, and change the value field to DEN1950, near the bottom of the list of attributes).
How does the pattern of largest values in the DEN1950 map differ from that of the POP1950 map you made a moment ago? Answer: _____

_____

Is the value range still a problem? Answer:_____

Why? Answer:_____

_____

Let's try a different attribute. This time use BRTHP000, which is the number of births per 1,000 people.

This variable should offer a more visually interesting map, with greater variation among the countries. Which continent has the highest birthrates? Answer: _____

Which countries beyond this continent have high birthrates? If you're not sure of their names, use the Identify tool to find out. Answer:_____

❶

Identify

Some of the countries may be too crowded to click easily. In this case, it will help to zoom in on those countries.

Zoom In

After zooming in, you can zoom out again (using the negative zoom tool), or you can use the Pan tool to pan across the map, or you can zoom to the extent of the entire active layer.

Zoom Out

Pan

Using MapTips

As long as you're practicing identifying country names, there's an even quicker way to check them: MapTips. Click on the MapTips icon:

MapTips

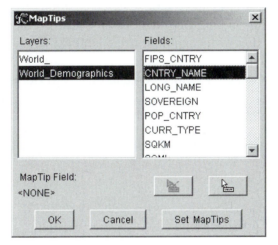

This icon brings up the MapTips window. Since you want to quickly identify Country Names (Attribute, or Field, = CNTRY_NAME) and the country name information is in the **World_Demographics** layer, you can choose these variables in the MapTips window. (Remember, the **World** layer is just a background layer.)

Once you have set these variables, click Set MapTips and then OK.

Now try rolling your mouse pointer over the map. Does this help you identify country names faster? Answer:_____

Identifying country names is helpful, but it's probably not much of a challenge for you. So let's try setting a different MapTips field. If you still have your map showing birthrates, it might be nice to see the actual numbers as well as the 5 (or so) classes. Try setting your MapTips to show BRTHP000.

Now zoom in on Asia. As you move your cursor around, you should see numbers that allow you to compare countries in more detail.

Zoom In

Look at Burma (Myanmar) as you work. It shows a birthrate of 0. This is really low and probably not realistic! The 0 value most likely represents a lack of data. Burma may not have kept up-to-date in its reporting of statistics in recent years. Occasional missing values are a problem that all good map makers need to keep in mind.

Now try setting your MapTips to DETHP000. As you might guess, this represents death rates. As you move your cursor around, do you find a generally positive correlation between birthrates and death rates, or a negative correlation? That is, in the high birthrate countries, do you find a high death rate as well, or a low death rate? Answer:_____

_____

Can you make global generalizations about the relationship of these two variables from perusing a small part of the world map? The answer is probably yes and no. You might propose possible patterns, and even guess cautiously at explanations. It is always best to look more carefully at the big picture before making broad generalizations, however. Statistics, graphs, or maps of correlation coefficients are possible methods to gain confidence in the patterns that you believe you see.

Map Variables of Your Choice
The **World_Demographics** shape file has a number of different attributes available. You'll find descriptions of them in the Appendix at the end of this exercise. Try mapping one or two additional attributes. For each, pause to look at your map, both globally, and by zooming in on regions. Can you identify countries with high and low values of your selected attributes? Can you propose some explanations of those highs and lows?

Additional Tools in Your Toolbar
There are a few other tools you should try before you end this exercise. One is the measuring tool:

Select the measuring tool, specify Miles as measuring units, then measure the distance across the United States. What distance do you get? Answer:_____

Measure

Does this seem like a reasonable distance? Answer:_____

If your measuring tool reports a distance greater than about 3,000 miles, then ArcExplorer is using the wrong measurement system—units different from those defined in the data file you are using. You need to reset your scale bar properties. You can do this by going to the View menu > Scale Bar Properties > Map Units, then selecting Meters. Now ArcExplorer will know what units to use in calculating distances.

Now measure the width of the United States. Do you get a distance closer to 3,000 miles? _____

Try measuring the distance in kilometers. How wide is the country in kilometers?
Answer:_____

Note: Anytime you use the measuring tool, check that your results seem to be in the right range, and if they do not, then check the Scale Bar Properties as you have just done.

Another useful tool is the Layer Properties tool. You have also opened the Properties window by clicking on the layer name, pressing Ctrl-L, or using the Layer menu.

Layer
Properties

The Save icon (or File menu > Save Project) is important if you wish to return to your project later. Since you copied this project from a CD, it is a read-only file. That is, you cannot save changes to it. Instead, save your project with a new name. Go to the File menu > Save As, and call your project **World_Demographics2.axl.** (Or if you are working in a public computer lab, give it a name you'll recognize as your own.)

If you have created a new project, you can simply click the Save Project icon to save your work. Note that ArcExplorer will automatically give your project the ".axl" extension to indicate that the file is an ArcExplorer file.

Save
Project

The .axl file you have just created does NOT *contain* all the maps and data you have worked with. It is a simple list of *instructions* that tells ArcExplorer what shapefiles to use, where to find them, what colors to use in displaying them, and at what scale to display them. You can come back to ArcExplorer and open the same .axl file (using the File menu > Open File, or the Open File icon at the left of the toolbar).

If you move your .axl file, perhaps putting it on a floppy disk to take home, it will NOT open properly. The actual data reside in the shapefiles. The project (.axl) file only contains instructions for displaying them.

Create an Image to Print
Before you finish, you should print a map. There are two ways to do this. One is to click on the Print tool, which will find your computer's printers and print your map. Another is to use the Copy Map to Image File tool:

Copy Map
Image to File

For now, try using the Copy Map to Image File tool. You will be asked to supply a file name, and the dialog box will show which folder or directory in which the image file will be saved. If you just saved your project, the file should be put in the same folder by default. The file format will be JPG (.jpg), the standard format used in web pages and many other applications.

Make and Explain Your Own Map
Set your map to show another variable of your choice, using Graduated Symbols. Then save your map to an image file.

Once you have created your image file, open a word processing program, such as Word. Insert your new image. (In Word, use the Insert menu > Picture > From File, and navigate to the directory where you just saved your image and then select that new image.)

Above your map, in your word document, type a title. Below your map, type a short paragraph explaining what variable you mapped, and identifying a country or region with especially high values and a country or region with especially low values.

What Have You Learned?
You have accomplished a great deal in this exercise. You added and removed layers, changed symbols from single to unique to graduated classifications, and inspected a number of different attributes. You rearranged the display order of your layers. You changed the colors and color ranges used to display the data. You learned to select features, either singly or in groups. You used the information tool, the Attributes tool, and MapTips to investigate descriptive attributes for different countries. You learned to move around the map and zoom in and out. You learned to save your project, and to create an image of your map that could be imported to another program.

In other exercises, you'll use these skills to practice buffering features, selecting by using the query tool, extracting tabular information, and other tasks. We hope you enjoy your exploration into GIS!

Appendix to Exercise 1: Data Dictionary

Like most GIS programs, ArcExplorer requires short attribute names. As a result, attribute names (field names) are often cryptic. Below are explanations of attributes associated with countries in World_Demographics.shp. All data not dated are most recently available records or estimates, as of 2003. Data source: US Census IDB, 2003.

Item Name	Definition
FIPS_CNTRY	Country identification code
CNTRY_NAME	Name of country
POP1950	Population 1950
POP1960	Population 1960
POP1970	Population 1970
POP1980	Population 1980
POP1990	Population 1990
POP2000	Population 2000
POP2010	Population 2010 (projected, as of 2003)
POP2020	Population 2020 (projected, as of 2003)
BRTHP000	Births per 1,000 people
DETHP000	Deaths per 1,000 people
MIGRP000	Migrants per 1,000 people
NETINPCT	Net increase, percent per year, including births, deaths, immigration, and emigration
BIRTHS	Number of births per year
DEATHS	Number of deaths per year
NUMMIGR	Number of migrants per year
NATINCRS	Natural increase (births, deaths alone)
POPCHANG	Population change, in number of people
FERTPWOM	Fertility per adult woman
IMRBOTH	Infant mortality rate, male and female
IMRMALE	Infant mortality rate, male
IMRFEMALE	Infant mortality rate, female
LIEXBOTH	Life expectancy, male and female
LIEXMA	Life expectancy, male
LIEXFEM	Life expectancy, female
DEN1950	Density, persons per square mile, 1950
DEN1980	Density, persons per square mile, 1980
DEN2000	Density, persons per square mile, 2000
DEN2020	Density, persons per square mile, 2020
SQKM	Area, square km
SQMI	Area, square miles

Use the information in the Attributes window to answer the following questions about this neighborhood.

What was the population of the block in 2000? Answer: _____

How many Hispanic people live on the block? Answer: _____

How many females live on the block? Answer: _____

How many people on the block are under 18? Answer: _____

How many people on the block are over 65? Answer: _____

How many families live on the block? Answer: _____

Are there any vacant houses? If so, how many? Answer: _____

Do most of the residents own or rent their homes? Answer: _____

Close the Attributes window.

Geography of the Neighborhood

Four streets border the neighborhood block. Use the Identify tool to determine their names. In order to do so, first make active the road layer by clicking it in the legend. When a box appears around it, you know it is active. Also, check the box for the road layer in order to turn it on. Next, click on the Identify tool in the toolbar and click on each of the four roads surrounding the block.

Identify

What are the names of the four roads? (Hint: There are three avenues and one street.)
Answer: _____
Answer: _____
Answer: _____
Answer: _____

Which road is the longest? Answer: _____

What is the zip code of this block? Answer: _____

Close the Identify window.

Locating Neighborhood Landmarks

Landmarks are often helpful when it comes to giving and following directions. You can use the buffer tool to locate landmarks located within a few miles of your block. Making sure that your block is still highlighted in yellow, look at the legend and make active the

Buffer

Brooklyn_blockdem layer. Also make sure that it is turned on (the box is checked). Click on the Buffer button on the toolbar.

Choose a buffer distance of two miles and select features from the **Brooklyn_landmarks** layer.

```
Buffer                                        [X]

    Buffer Distance:    [2      ]

    Buffer Units:       [Miles         ▼]

    ☑ Use buffer to select features from this layer.
    [brooklyn_landmarks       ▼]

                   [   Clear Buffer   ]

        [  OK  ]   [ Close ]   [ Apply ]
```

Click OK. Your entire map becomes red. Click on the Zoom Out tool in the toolbar. Using that tool, click on the map until you see the red circle that indicates the two-mile buffer around your block. Use the Pan tool to center the red circle in the map window. If your circle appears too small, use the Zoom In tool to enlarge it.

🔍
Zoom Out

🖑
Pan

🔍
Zoom In

Activate the **Brooklyn_landmarks** layer and turn it on by checking the box in the legend. As you can see, all landmarks within the two-mile buffer zone are now labeled and highlighted in yellow.

How many landmarks are within the buffer zone? Answer: _____

You can change the appearance of the labels for the landmarks by using the Layer Properties button. Click the Layer Properties button to open the dialogue box. In the drop-down menu choose the Labels tab. Change the font size of the NAME labels to at least 16 and click OK.

Layer Properties

```
brooklyn_landmarks Properties                    [X]

  Symbols  Labels  General
  ┌─ Label features using: ─────────────────┐
  │                                          │
  │      [NAME                    ▼]         │
  │                                          │
  └──────────────────────────────────────────┘

    Font [Arial                      ▼]

    Size [16 ↕]  ☑ Bold    ☐ Italic

        [ ◀ Effects ]

    Color [■ Black    ▼]    [ Preview ]




              [  OK  ]  [ Cancel ]  [ Apply ]
```

change, slope *steepness* was determined. In the image you have just turned on, darker colors indicate steeper slopes. Where are slopes steepest in this image?
Answer:_____

To quantify slope steepness, turn on the layer **contour100m** and drag it to the top of the legend so it is visible over the **slopes.tif** layer. This file shows contour lines, or lines of equal elevation, at 100 m intervals. Once again, these contour lines were calculated from the DEM: lines were drawn connecting cells of equal elevation value.

You can quickly see steep areas because the contour lines crowd together over a small horizontal distance. Generally, where do you see the steepest slopes?
Answer:_____

Zoom in on the caldera, and count the contour lines in the steepest part of the inner rim. If each line represents 100 m of elevation change, how high are the steep slopes of the lake? Answer: _____

Zoom In

Now use your measuring tool to calculate slope steepness. Remember that when you use the measuring tool, you must first set the Scale Bar Properties, in this case to meters (View menu > Scale Bar Properties > Map Units > Meters).

Measure

Once your properties are set, use the measuring tool to measure the horizontal distance of the steepest part of the crater's interior rim, in meters. Determine the steepest slope you can find by measuring vertical rise and horizontal distance and then dividing.

Vertical rise (number of contour lines x 100 m):_____meters

Horizontal distance: _____ meters

Slope: _____

Now zoom in on the river valley in the southwest corner of the map. If you follow the pointed contour lines from the map margin up to the beginning of the sharply incised valley, approximately how great is the elevation change? _____

Zoom In

What is the horizontal distance over the contours you just counted, in kilometers?
Answer:_____

How fast does this stream drop, in meters per km? Answer:_____

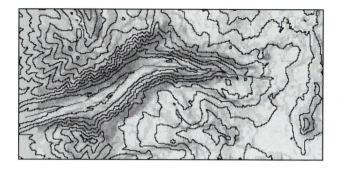

Examining Roads on Slopes

Understanding topography is important for many reasons. For example, in areas under development, road building on steep slopes can cause erosion, reduce road stability, and increase deposition of sediment into stream channels. Sediment in streams degrades habitat for fish and other aquatic organisms.

Now that you're familiar with slopes in the Crater Lake area, look at the distribution of roads. Turn on the **roads** layer. For better visibility, turn off the 100 m contour lines. Also turn off the **slopes.tif, hillshade.tif,** and the **DEM.tif** layers, to speed up the pace at which the program redraws the map.

As you look at the roads, can you see where the park boundary is? (Describe the location.) Answer:_____

To check whether you were right in guessing the park boundary, find the layer **park_boundary,** and turn it on. What is the effect of the boundary line on road construction?
Answer: _____

You can distinguish the major, named roads and highways by turning on the layer **major_roads.** These roads are extracted from the larger set of roads. Turn on the MapTips for ROADNAME, and roll your curser over the major roads to see some of their names.

MapTips

As you might guess from their names, the major roads are access roads to the park and the caldera.

Now look again at the **roads** layer. Change the symbols for roads to Unique Values, using ROADTYPE as the classification field. Set the "Unimproved, dirt" road type to red. Note that the roads displayed in this layer are broken into small segments. Each time a road branches, a new segment begins, with its own length and steepness characteristics.

Layer
Properties

Are most of the *unimproved dirt* roads in the park or outside of the park?
Answer:_____

Propose a plausible explanation for the distribution and density of unimproved, dirt roads in your map. What might be the purpose of these roads?
Answer:_____

To investigate further the purpose and effect of the unimproved dirt roads, add the image layer **treesize.tif.** Make **treesize.tif** visible. This layer provides a generalized classification of tree maturity within the park. (Similar data outside the park, unfortunately, are not readily available.) Colors grade from brown (no significant tree cover) to blue (mature trees).

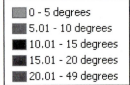

Drag the **roads** layer to the top of your legend and make sure it is visible. Notice the regions of small or regrowing forest near the unimproved roads? What might these areas represent? Answer: _____

Does the park boundary entirely protect forests from logging? Answer:_____

There are other areas of absent or regenerating forest within the park boundaries, including blocks to the north and east of the caldera. To inspect these, set the transparency of the **treesize** layer to about 70 percent (double click on the layer name to open the properties window, then set transparency using the slider). Then zoom in on the area around the lake to examine yellow patches.

Zoom In

How might you explain the large, treeless blocks around the lake's margin?
Answer:_____

Slope Steepness

Among the road classes in the roads layer, unimproved, dirt roads are most susceptible to further erosion because their surfaces are relatively soft. Erosion risk is most serious, however, on steep slopes. You can evaluate whether the unimproved dirt roads are mainly on flat or steep slopes. To do so, add the image layer called **roadslope.tif** using the Add Layers button. This layer shows slopes within 100 m of roads classified into five slope classes, light blue to purple, as shown at right.

Add
Layers

Look over the **roadslope** layer, to assess visually whether most roads are on steep slopes (purple shades) or flat areas (light blues).

Drag the **roads** layer to the top of the legend. Now select the Secondary Highways (in the **roads** layer, click on the line symbol next to the Secondary Highways). The Secondary Highways will turn bright yellow when highlighted. Do these highways occupy much of the purple area? Answer: _____

Clear the selection by clicking on the Clear All Selection tool.

Clear All
Selection

Examine the slopes and the different road types visually. Based on your visual estimate, which road class would you say is most likely to occur on steep slopes?
Answer:_____

Now let's quantify the slopes of these roads. Display the roads using Graduated Colors, with Avg_Slope_ as the value field. This field shows the average slope angle that each road segment crosses, in degrees (0 degrees represents a flat surface, 90 degrees would be vertical). A road might have a fairly level surface, but if it crosses a steep hillside, it tends to destabilize the slopes above. Where roads cut across slopes steeper than 15 or 20 degrees, sediment is more likely to wash off the hillsides—and into streams—in a storm.

As you look at your roads displayed by slope class, where would you say the roads are most likely to cross steep slopes? Answer:_____

Although we know that there are many unpaved, unimproved roads in this area, we have not yet assessed how extensive these roads are, or how steep. To answer these questions, use the Query Builder tool.

Query
Builder

Make the **roads** layer active. Then use the Query Builder tool to find roads whose ROADTYPE is Secondary Highway:

Note: When asked if you wish to display all records, indicate YES.

Once you have displayed all secondary highways, click the Statistics button to find the total length, in kilometers. Make sure you check the Use Query Results? box, or you'll simply summarize the entire table. Here you only want secondary highways. Then click OK to see your results.

Find out how many roads are within 50 m of this river: while the Rogue River is still selected, and the **major_streams** layer is still active, click on the Buffer tool. Set the buffer to 50 meters, and use it to select features from the roads layer:

Buffer

Buffer

Buffer Distance: 50

Buffer Units: Meters

☑ Use buffer to select features from this layer.

roads

Clear Buffer

OK   Close   Apply

Click Apply to create the buffer and select roads. Make sure the **roads** layer is visible. This process may take a few moments, so be patient. After applying the buffer, look at the roads selected by the buffer. Can you count them?

If not, then make the **roads** layer active. Then click on the Attributes tool to see a table of all the selected roads.

Attributes

How many road segments are within 50 m of this river? Answer:_____

Attributes

15 features

NewFeature1		Field	Value
NewFeature2		ROADTYPE	Unimproved, dirt
NewFeature3		ROADNUMB	FS 760
NewFeature4		ROADNAME	
NewFeature5		LENGTH_KM	6.4272
NewFeature6		LENGTH_MIL	4.017
		AVG_SLOPE_	7.5

Layer: roads

Of the roads within 50 m of the Rogue River, focus on the unimproved, dirt ones. Click through the list, and for each unimproved, dirt road, record in the table below the length (in km) and average slope. Also find the sum of lengths for each stream.

Rogue River		Red Blanket Creek		Castle Creek	
Length	Average slope	Length	Average slope	Length	Average slope
Sum of lengths:		Sum of lengths:		Sum of lengths:	

To fill in this table for the other two streams, make **major_streams** layer active again, and repeat the Find, Buffer, and Attributes steps for Red Blanket Creek. Then do the same for Castle Creek (do not use Little Castle Creek).

Of the three rivers, which one is most likely to be susceptible to erosion and sedimentation, based on the cumulative length of unimproved, dirt roads you have found? _____

Which stream has roads on the steepest slopes? _____

While one of your streams is still showing the 50 m buffer, zoom in on the buffered area. Do you think that 50 m is a good distance to use? Would it have made a great difference if you had used 100 m?
Answer:_____

_____

Erosion into these streams depends on a variety of factors, but the methods you used here are a good start at comparing vulnerability of streams. What other variables would be important if you were to carry out a complete assessment of stream vulnerability in this area?
Answer:_____

_____

Zoom In

What Have You Learned?
In this exercise you have performed many tasks involving terrain analysis and comparing erodibility on slopes. You began by inspecting the shape of the land surface using a shaded digital elevation model (DEM). Then you added hill shading to give a

visual impression of shadowing on the landscape, and you adjusted transparency of the DEM to show the two layers together. You inspected slope steepness visually by displaying an image of calculated steepness values, and you quantified steepness by counting contour lines and measuring horizontal distance, both on the steep slopes of the caldera and along the run of Red Blanket Creek.

You have inspected roads in this area, both by surface type and by steepness of slopes along which roads are built. You have examined vegetative cover and proposed some explanations for differences in tree size as a function of topography and roads in the area. You have used the Query tool to identify roads of different surface types and to produce summary statistics for length and slope steepness of each road class. You may have found that your quantitative results varied somewhat from your initial visual estimates.

Finally, you have selected three streams, buffered them, and used buffers to identify roads within 50 m of these streams. By focusing on the most erodible surfaces (unimproved, dirt roads) you have produced a preliminary comparison of erosion vulnerability among these three streams.

The skills of visual assessment, measurement, querying, and tabulating statistics for selected features are all central to GIS analysis. This exercise has provided some basic examples of how you can evaluate environmental conditions and identify vulnerable areas using these tools.

Exercise 4: Earthquake Histories and Hazards

Earthquakes are one of the most important geological hazards affecting human populations. Where do earthquakes occur, and why do they happen where they do? How can we estimate the vulnerability of large populations to these unpredictable events? Here we'll explore the distribution and risks of earthquakes around the world.

Getting Started
Before starting this exercise, copy the directory **Earthquake_exercise** from your CD to the C:\ExploringGIS\ folder on your computer. Then open ArcExplorer. Open the project called **Earthquake_exercise,** which is in the folder you just copied from the CD. Do this by either using the Open Project button on the toolbar or by going to File > Open Project (Control O) and then navigating to the earthquake folder.

Open Project

The map that you have opened shows the locations of worldwide earthquake epicenters from 1973 to 1999 symbolized with a green triangle. There are 7,787 earthquake records in the database used to draw this map. The map also displays the configurations of the earth's tectonic plates. Along nearly all lengths of tectonic plate boundaries earthquakes have occurred. Therefore, earthquakes, even more than volcanoes, help to define and delineate plate boundaries.

Earthquakes and Plate Tectonics
In the left-hand column of the ArcExplorer window, called the Legend, you will see the layer files for this exercise with a small square next to each one. The **world_quakes** and **plates** layers are checked and are viewable in the map window.

Though there is no scale bar on this map, you can easily add one. To insert a scale bar, go to Scale Bar under the View menu (as shown at right).

You should now have X and Y coordinates, a ratio scale, and a scale bar at the bottom of your map. Move the cursor around the map surface and notice that the X (latitude) and Y (longitude) values change with the location of the cursor. The ratio scale indicates that one unit on the map represents some larger number of units on the earth's surface. Those units may be feet, miles, meters, or kilometers. The scale bar shows the distance on the earth to which one inch on the map is equivalent. The map, scale, and screen units can be changed by choosing Scale Bar Properties under the View menu as shown below:

Earthquake Magnitude

Let's explore the earthquake data. Single click on the **world_quakes** file. Then click on the Query Builder button. You will open the window that is shown below:

```
Query Builder                                    [x]

Select a field:                          Values:
YEAR
MO              [ < ] [ = ] [ > ]
DA
HOURS           [<=] [<>] [>=]
MINUTES         [and][ or][not]
SECONDS
DEP             [ % ][like][ () ]
MAG

[0                                                ]

   [ Execute ]   [ Clear ]

[✓] Show All Attributes   Display Field: [YEAR   ▼]



Query Results:
[Highlight][ Pan ][ Zoom ][Statistics][ ■ ]
```

On the left-hand side of the Query Builder window are the fields of data available for each earthquake. Many of the fields are abbreviated but are self-explanatory (i.e., MO is short for month and DA is short for day). We can locate an earthquake and find out what year, month, day, and time it occurred. We can also learn the depth (in kilometers) and magnitude of the earthquake. Let's look at the magnitude field in more detail.

Single click on the MAG (or magnitude) field. Your Query Builder window will be updated with values of all the magnitudes of the recorded earthquakes in this dataset. In the Values column on the right-hand side, you will see earthquake magnitudes ranging from 5.0 to 8.3. Scroll down to see those values. If you look at the text box to the right, you'll see that these are all moderate to great earthquakes. Obviously, some earthquakes have been left out of the dataset because we know that earthquakes can be detected lower than M5.0. To find out how many M8.3 earthquakes have occurred between 1973 and 1999, let's query the data. First, in the Query Builder window, use the Clear button to start the query dialog. Single click MAG in the Field column, then single click the equals sign (=), and then single click 8.3 in the Values column. Your query should look like that shown in the oval below or (MAG = 8.3). Now click the Execute button. You will see listed two

What is M? M stands for magnitude, which is the most common measure of an earthquake's size. An incremental increase in magnitude by one unit (e.g., 4 to 5 or 5 to 6) means the ground moves 10 times faster. An informal classification of earthquake magnitude shows how seismologists describe an earthquake:

Under 5	small
5-6	moderate
6-7	large
7-7.8	major
7.8 and above	great

earthquakes with a magnitude of 8.3 in the records. These results are summarized in the lower left corner of the Query Builder window. Also note that your map now has two earthquakes highlighted in yellow (shown below in the squares).

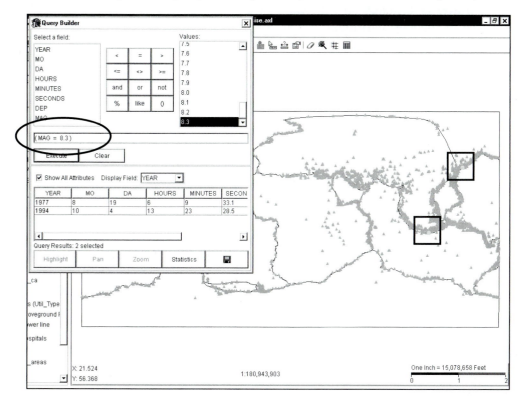

To zoom to both of these earthquakes, in the lower half of the Query Builder window click once on the '1977' line, then hold down the shift key and click once on the 1994 line. Now click the Zoom button in Query Builder (not the Zoom In button in the toolbar!) and you will be zoomed into the two largest earthquakes in the dataset. Move but don't close the Query Builder box to see the earthquakes. Zoom back out by using the Previous Extent button in the map window (not in the Query Builder window) and then click the Zoom to Full Extent button. Click the Clear button in the Query Builder window.

Previous Extent

Zoom to Full Extent

How many earthquakes of magnitude 8.2 are there in the dataset?
Answer: _____

Find the deepest earthquake in your query of M8.2 quakes. Click once on it, then use the Pan button in the Query Builder window to reveal the deeper of the two earthquakes in this dataset.

On what plate is this deep earthquake located?
Answer: _____

Identify

(Hint: Click once on the **plates** layer, click on the Identify tool button, and then click once on the plate that the deepest earthquake is on. You can also use the Zoom In tool, if you need to zoom in closer.)

Zoom In

What are the approximate X and Y coordinates of the earthquake? To determine them, rest the cursor directly over the highlighted deep earthquake.
Answer: X (longitude) = _____ and Y (latitude) = _____

How many earthquakes of magnitude 5.0 are there in the dataset? (There are more than 100, so you will need to click Yes when asked to see the records.)
Answer: _____

Next, find the deepest earthquake in your query of M5.0 quakes. To do this, click the Statistics button in the Query Builder window. Select DEP from the pop-up window, check the box in the pop-up window that asks Use Query Results?, and click OK. The Statistics Results reveal that the deepest earthquake of M5.0 occurred at 619 kilometers depth. To determine which of these M5.0 earthquakes occurred at 619 kilometers depth, build another query in which DEP=619. Of the two earthquakes in the dataset that occurred at 619 kilometers, only one was a M5.0 earthquake. Click on that earthquake which is listed in the lower half of the Query Builder window and use the Pan button at the bottom of the Query Builder to reveal its location on the map.

What are the approximate X and Y coordinates of the earthquake?
Answer: X (latitude) = _____ and Y (longitude) = _____

Next, make the plates layer active and use the Identify tool to determine the plate on which that earthquake occurred. List the name of the plate here.
Answer: _____

Close the Query Builder window and click Zoom to Full Extent.

Zoom to Full Extent

Earthquake Timing

Do earthquakes occur at a typical time of day? Let's find out. Make sure that the **world_quakes** layer is active and use Query Builder again. Click Clear to start a new query. Begin a query of the hour that earthquakes occur by clicking once on HOURS in the Field column. You will see numbers from zero to 23; these values indicate an hour in a day where zero is the midnight hour or the 24th hour, 1 is 1:00 a.m., 13 is 1:00 p.m., 23 is 11:00 p.m., and so on. You will query each hour and write down how many earthquakes occur during each given hour. To get you started, click once on HOURS, click once on the '=' sign and click once on 0, then click Execute. There are over 100 records, so click Yes to see them. Write

Query Builder

down the Query Results number below for the zero hour. Clear the query and then proceed to do the same for the remaining hours.

0 hour = _____	hour 6 = _____	hour 12 = _____	hour 18 = _____
hour 1 = _____	hour 7 = _____	hour 13 = _____	hour 19 = _____
hour 2 = _____	hour 8 = _____	hour 14 = _____	hour 20 = _____
hour 3 = _____	hour 9 = _____	hour 15 = _____	hour 21 = _____
hour 4 = _____	hour 10 = _____	hour 16 = _____	hour 22 = _____
hour 5 = _____	hour 11 = _____	hour 17 = _____	hour 23 = _____

The raw numbers can be deceiving. Let's graph these numbers and see if there is a more typical hour when an earthquake might occur. Use Excel or some other graphing program to plot the data that you've culled from the earthquake database and make a bar graph.

The following is a description of how to create a bar graph in Microsoft Excel. Your instructor can explain another software program or a bar graph can be easily drawn by hand.

Using an Excel spreadsheet, in cell A1 type 'Hour of EQ.' In cell B1 type 'number of EQs.' In cell A2 type the number one and fill down the A column with the numbers of hours to number 24. You will use the zero hour data for the number 24 in Excel. The setup for your spreadsheet should look like the figure below.

Beginning in cell B2, type in the number of earthquakes for each of the hours that you determined in the preceding queries and recorded above. Remember that the zero handwritten entry will be entered last under the 24 hour. Now go to Chart

Wizard on the toolbar and create a column-type graph. Click through using the default button and study your graph. Numbers on the x-axis are the hours at which earthquakes occurred and y-axis values are the numbers of earthquakes. Each bar on the graph indicates the number of earthquakes that occurred at a given hour. Use the graph to answer the following questions.

At what time of day have the most number of earthquakes occurred?
Answer: _____

At what time of day have the fewest number of earthquakes occurred?
Answer: _____

If you have printing capability or if you drew your graph, include it with your exercise.

Turn off the Query Builder window, click on Zoom to Full Extent, and click the Clear All Selection buttons.

Zoom to
Full Extent

Clear All
Selection

Earthquakes and Population
In this part of the exercise you will examine the earthquake-prone state of California in close detail and see how earthquakes and earthquake faults have the potential to affect people in that region of the United States. Keep the **world_quakes** layer on. Turn off the **plates** layer. Turn on the **ca_faults,** the **ca_counties,** and the **urban_areas** layers. Click once on the **ca_faults** layer to make it active and click the Zoom to Active Layer button. Displayed before you is a map showing the counties in California with labels and black lines, urban areas of California, shown by yellow polygons, and active or recently active earthquake faults in California, shown by purple lines.

Zoom to
Active Layer

Notice the locations of the urban areas shown in yellow. These are regions of significant population density in urban, as opposed to rural, areas of the state. These areas do not necessarily correspond to county line boundaries; rather they tend to cross those boundaries.

There are four urban areas that stand out in this layer. The urban areas may be more noticeable if you click off and on the **world_quakes** layer. Starting from the south and proceeding north, one encounters the San Diego Metropolitan Area, the Los Angeles Metropolitan Area, the urban areas that comprise the San Jose Metropolitan Area, Oakland, Alameda, Union City, Fremont, Concord, and San Francisco, all of which are referred to as the Bay Area, and Sacramento. In this portion of the exercise, you will compare these four urban areas by looking at population, the number of faults in each vicinity, and the number of earthquakes that have occurred in each region.

To make the earthquake locations more visible, change the symbols for the **world_quakes** dataset. Double click on the **world_quakes** layer and you will get the

Layer Properties dialog box. Under the Symbols tab change the 'Draw features using' to Graduated Symbols. As shown in the figure below, use the MAG field, use three classes, make the style of the symbol a triangle, change the colors to start with green and end with red, and then change the size of the symbol to start with 9 and end with 15. Click OK to update your map of California. Notice that the earthquake symbols now show not only location but earthquake magnitude.

world_quakes Properties			☒

Symbols | Labels | General

Draw features using:

Graduated Symbols ▼

Field   MAG ▼

Classes 3 ▲▼   Style Triangle ▼

Color

Start ■ Green ▼

End ■ Red ▼

Size

Start 9 ▲▼

End 15 ▲▼

Symbol	Range	Label
▬	Less than 6.1	Less than 6.1
▬	6.1 - 7.2	6.1 - 7.2
▬	7.2 - 8.3	7.2 - 8.3

OK | Cancel | Apply

Begin your urban analyses in the San Diego Metropolitan Area (the southernmost yellow area on your map of California). The procedures we use for the San Diego Metropolitan Area you will also use for the other three urban areas.

Population Analysis
Using the Zoom In tool, drag a rectangle around the urban area of San Diego. Click once on the **urban_areas** layer to make it active and use the Identify tool to click on the yellow polygon for the San Diego Metropolitan Area. The Identify Results window will open and list the population for the urban area.

Zoom In

Identify

Population of San Diego Metropolitan Area
Answer: _____

Now make the **ca_counties** layer active. Your Identify tool should still be active so click once somewhere in the white region of San Diego County. Please note that if you do not activate the correct layer, you will get incorrect answers. Shown below, the Identify Results box on the left is for the **urban_areas** layer, which is circled for your reference, and the Identify Results box on the right is for the **ca_counties** layer.

Urban_areas layer   **ca_counties** layer

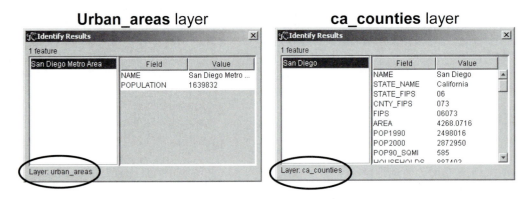

Read the population value from the POP2000 field on the right-hand side of the Identify Results box for the **ca_counties** layer and record that value.

Population (2000) of San Diego County
Answer: _____

As you can see from this example, the San Diego Metropolitan Area has a lower population than San Diego County and occurs wholly within that county. However, as stated above, urban areas do not necessarily fall neatly within a single county. To illustrate this point, use the Zoom In, Zoom Out, and Identify tools and alternately activate the **urban_areas** layer and the **ca_counties** layer, to collect population data for the urban areas and counties in the more northerly metropolitan areas of California listed below.

Population (2000) of
    Los Angeles County   Answer: _____
    Orange County   Answer: _____
    Los Angeles Metropolitan Area   Answer: _____

    San Mateo County   Answer: _____
    Santa Clara County   Answer: _____
    San Francisco County   Answer: _____
    Alameda County   Answer: _____
    San Jose Metropolitan Area   Answer: _____
    Oakland (urban area)   Answer: _____
    Alameda (urban area)   Answer: _____
    Union City (urban area)   Answer: _____
    Fremont (urban area)   Answer: _____
    Concord (urban area)   Answer: _____

San Francisco (urban area)	Answer: _____

Sacramento County	Answer: _____
Yolo County	Answer: _____
Placer County	Answer: _____
Sacramento Metropolitan Area	Answer: _____

Fault Segment Analysis

In this section of the exercise we will determine the number of active or recently active faults that exist in or adjacent to an urban area. To do this you will activate the **urban_areas** layer, select the polygon, and create a buffer around the urban area polygon. Then you will count the number of fault segments that occur within the urban area.

In the San Diego Metropolitan Area, click once on the **urban_areas** layer to make it active, then choose the rectangle under the Select Features tool. Click once in the San Diego Metro Area to select it; notice that you don't need to draw a rectangle around the polygon. Once the urban area is selected, a yellow line will be drawn around the perimeter. If the urban area does not change when selected, the layer is not active. Open the Buffer window (as shown below) by clicking on the Buffer tool. Now use a buffer distance of 10 miles, select features from the **ca_faults** layer, and click OK.

Select Features

Buffer

Buffer

Buffer Distance: 10

Buffer Units: Miles

☑ Use buffer to select features from this layer.

ca_faults

Clear Buffer

OK   Close   Apply

A buffer will be drawn in red with a 10-mile radius around the San Diego urban area; faults within that buffer will be highlighted in yellow. To determine the number and nature of these faults, activate the **ca_faults** layer by clicking on it one time and then choose the Attributes tool. The Attributes window will open and will indicate the number of fault segments that exist within 10 miles of the San Diego urban area. Four faults occur in that zone, and they are well located rather than concealed, approximately located, or inferred. When you have finished examining the San Diego Metropolitan Area, use the Clear All Selection tool to begin to examine a new urban area.

Attributes

Clear All
Selection

In order to analyze multiple urban areas in one region such as the San Francisco Bay Area, hold down the shift key as you use the Select Features tool; a yellow line will be drawn around each of the polygons. If you make an error during the selection process, use the Clear All Selection tool to start again. Note too, that if the Attributes tool appears gray, this means that there are no faults within the specified buffer distance.

Select Features

How many active or recently active fault segments are located within 10 miles of the urban areas?

San Diego Metropolitan Area   Answer: _____

Los Angeles Metropolitan Area   Answer: _____

San Jose Metropolitan Area, Oakland, Alameda, Union City, Fremont, Concord, and San Francisco Bay Area   Answer: _____

Sacramento Metropolitan Area   Answer: _____

Earthquake Analysis
Large magnitude earthquakes in major urban areas sometimes cause significant losses of life and property. When buildings, such as hospitals and fire stations, and infrastructure, like pipelines and roadways, are damaged, the number of fatalities can increase. In this portion of the exercise, you will analyze the four urban areas in California in terms of the number of major earthquakes that have occurred in or near each area. In doing this portion of the exercise, use a buffer of 25 miles as this will serve as a conservative estimate of the distance between an earthquake that could cause substantial damage and the metropolitan area nearest to it. For example, the October 1989 M7.1 Loma Prieta earthquake caused major damage to the Mission District and the Bay Bridge in the San Francisco Bay Area, both well over 50 miles away from the epicenter of the earthquake.

Select Features

Buffer

Attributes

Clear All
Selection

Zoom in to the San Diego Metropolitan Area and activate the **urban_areas** layer. Using the Select Features rectangle tool, click once on the San Diego polygon. The polygon turns yellow indicating that you have selected it. Next, using the Buffer tool, choose a buffer distance of 25 miles and select features from the **world_quakes** layer. Click OK and observe the limits of the red buffer that has been drawn on your map. Activate the **world_quakes** layer. Notice that the Attributes tool is gray and cannot be utilized. This means there are no recorded major earthquakes within the specified buffer distance of the San Diego Metropolitan Area. Use the Clear All Selection tool to begin your examination of the other three California Metropolitan areas.

How many M5.0 or higher earthquakes have occurred within 25 miles of the following urban areas between 1973 and 1999?

San Diego Metropolitan Area   Answer: _____

Los Angeles Metropolitan Area   Answer: _____

San Jose Metropolitan Area, Oakland, Alameda, Union City, Fremont, Concord, and San Francisco Bay Area   Answer: _____

Sacramento Metropolitan Area   Answer: _____

Clear All Selection

Click the Clear All Selection button and click once on the **ca_faults** layer to make it active. Click the Zoom to Active Layer button and you will once again be able to see the entirety of the state of California.

Zoom to Active Layer

Earthquakes and Infrastructure

For the final part of this exercise you will study earthquake fault locations and their proximities to interstate roadways and hospitals. It is critical for the safety of people living in seismically active urban areas, that roads and hospitals be able to withstand the shaking that occurs with the earth ruptures. How well do various California urban areas fare with regard to the situation of roadways and hospitals?

If it is not active already, activate the **ca_faults** layer. Also, turn on the **ca_interstate** and **ca_hospitals** layers and move them so that they are above the **ca_faults** in the legend. Begin in the San Diego Metropolitan Area and work your way north.

To determine the number of hospitals that might be damaged during a large earthquake in the San Diego area, you will select active or recently active faults in the area and then apply a 500-foot buffer to those faults with the software selecting features from the **ca_hospitals** layer.

Select Features

First, click once on the **ca_faults** layer to make it active. Next, click on the Select Features tool and choose the polygon, not the rectangle as we have most recently, as your selection tool. Next, click a polygon shape around the perimeter of the San Diego Metropolitan Area. You need only click a short distance beyond the border of the metropolitan area (e.g., 5 miles). Once you have enclosed the San Diego Metropolitan Area in a polygon, double click to close the shape and have the area selected. Note that faults in the San Diego Metropolitan Area have turned from purple to yellow and therefore are now selected. If the faults were not selected, you likely had not activated the **ca_faults** layer. Once the faults in the urban area are highlighted, click on the Buffer tool. Choose a buffer distance of 500 feet, click on the check box and select features from the **ca_hospitals** layer. Now, click on the **ca_hospitals** layer and note that the Attributes too is gray. This indicates that no hospitals are located within the buffer limit. This is good news

Buffer

for the San Diego Metropolitan Area in the sense that no hospital is located within 500 feet of a fault.

▦

Attributes

Next, clear the buffer by clicking on the Clear All Selection tool. Using the same approach that you just used to locate hospitals in relation to faults, determine whether portions of interstate highways in the San Diego Metropolitan Area fall within 25 feet of an active or recently active fault. That is, once again select the faults, draw a 25-foot buffer around the faults, but for this example select features from the **ca_interstate** layer. Click on the **ca_interstate** layer to make it active and select the Attributes tool. As you can see, two portions of Interstate 5 (I5) come within 25 feet of an active or recently active fault in the San Diego Metropolitan Area.

✎

Clear All
Selection

How many hospitals exist within 500 feet of an active or recently active fault in the following metropolitan areas?

San Diego Metropolitan Area   Answer: _____

Los Angeles Metropolitan Area   Answer: _____

San Jose Metropolitan Area, Oakland, Alameda, Union City, Fremont, Concord, and San Francisco Bay Area   Answer: _____

Sacramento Metropolitan Area   Answer: _____

How many interstate highways in these California metropolitan areas are within 25 feet of an active or recently active fault?

San Diego Metropolitan Area   Answer: _____

Los Angeles Metropolitan Area   Answer: _____

San Jose Metropolitan Area, Oakland, Alameda, Union City, Fremont, Concord, and San Francisco Bay Area   Answer: _____

Sacramento Metropolitan Area   Answer: _____

Write an Earthquake Hazards Assessment Report
Consider the following hypothetical situation. The head of the corporation for which you work is considering whether or not to relocate her business to California. The company needs to move to an urban area where there will be a large number of people from which to draw a workforce. As the GIS consultant at your company, you have been asked to write a report evaluating the feasibility of the possible relocation based on your knowledge of earthquake hazards in the urban areas of California. Write a two-page report for the head of the company in

which you recommend one of the four metropolitan areas of California as the best area for the relocation. In your report, compare and contrast the four metropolitan areas in terms of: (1) population of the areas, (2) active or recently active faults within the areas, (3) earthquake history, and (4) hazards to infrastructure.

Close the project using the Close Project button or by exiting ArcExplorer. Do not save the changes.

Close
Project

What Have You Learned?
In this exercise you have learned to recognize tectonic plate boundaries using the locations of earthquakes. In the process you have learned a bit about earthquake magnitude and timing. Additionally, you have examined the relationships between locations of earthquakes and major population centers and considered the risks to people and infrastructure that earthquakes pose. With regard to GIS skills, you have had the opportunity to use the following tools: Query Builder, Zoom, Identify, Clear All Selection, Select Features, and Attributes. You have also learned how to construct a Buffer. Once again we see that GIS is a powerful tool for examining the relationship between people and our landscape.

Exercise 5: Spatial Patterns of Volcanoes and Volcanic Hazards

Getting Started

To begin, copy the folder **volcano_exercise** from your CD to your C:\ExploringGIS\ folder. Open ArcExplorer (Start button > ArcGIS> ArcExplorer). Then open the project called **Volcano_exercise,** in the folder you just copied to your computer. Do this by either using the Open Project button on the toolbar or by going to File > Open Project (Control O) and then navigating to the volcano folder.

Open Project

The map that you have opened shows dots on the locations of active volcanoes or volcanoes that have erupted in the recent past. It also displays the configurations of the earth's tectonic plates. The map is a geographic projection as opposed to an orthographic projection. That is, it shows the round earth as a flat surface.

Displaying Layers

In the left-hand column of the ArcExplorer window, called the Legend, you will see the project files for this exercise with a small square next to each one. The **volcanoes** and **plates** layers are checked and are viewable in the map window. Click on the check marks for the **volcanoes** layer and the **plates** layer to turn them on and off. If the box next to a layer is not checked, the layer will not be seen on the map.

Also notice that these files are truly layered. With the **volcanoes** layer and the **plates** layer both checked, click on the volcanoes layer in the Legend, hold it down with the mouse, and drag it so that it is underneath the **plates** layer. The **volcanoes** layer seems to disappear. If you drag it back up above the **plates** layer, it shows up again.

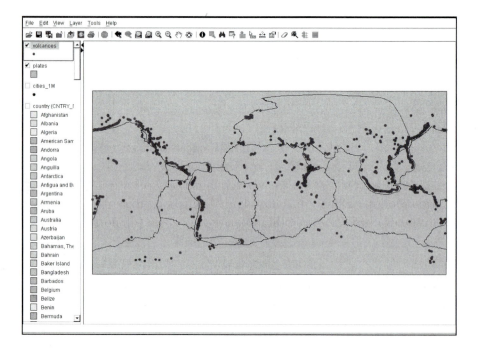

The ArcExplorer software randomly colors the volcano dots and the plate polygons. However, you can change the colors of these layers.

Left-click on the **volcanoes** layer (click on the word), then right-click on the same layer (again, click on the word). Choose Layer Properties. The Volcanoes Properties dialog box will open. (Alternatively, you can use the Layer Properties button on the toolbar or you can double left-click on the layer to open the Volcanoes Properties dialog box.) Under the Symbols tab, change the Style to Stars, the Color to Red, the Size to 7, and click OK. Your map will be updated with this change.

Layer
Properties

To see that discrete plates comprise the earth, we will color each of the plates a different color. Left-click once on the **plates** layer and open the Layer Properties dialog box. In the drop-down menu for "Draw features using" change the option to Unique Symbols. As shown below, in the "Field for values" choose PLATE_NAME and change the "Color Scheme" to Minerals. Click OK. Then, once again open the Layer Properties dialog box and click on the Labels tab. Label features using PLATE_NAME. Change the font to Arial and use bold to make it stand out. Click OK.

![plates Properties dialog box. Tabs: Symbols, Labels, General. Draw features using: Unique Symbols. Field for values: PLATE_NAME. Color Scheme: Minerals. Style: Solid fill. Remove Outline checkbox unchecked. Table with columns Symbol, Value, Label showing South American, Cocos, African, Scotia, Juan de Fuca, Pacific, Indo-Australian. Buttons: OK, Cancel, Apply.]

You should now have a map with red stars and fourteen named and colored tectonic plates.

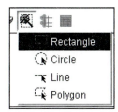
Select
Features

Attributes of the Volcanoes Dataset
How many volcanoes are in the dataset? To find out, first click once on the **volcanoes** layer to make it active. From the toolbar, click on the Select Features button and choose the Rectangle, as shown at right.

Now drag a rectangle around the outside of the map so that you include all of the volcanoes in your selection. Let go of the mouse. All of the red stars will change to yellow when they are selected. Use the Attributes button to open the Attributes window. This window will provide the names of all of the volcanoes in the dataset (in the left-hand column) and the attributes of each of the volcanoes. The number of features (the number of volcanoes selected) is shown under the word Attributes.

Attributes

How many volcanoes are shown? Answer: _____

Look at the right-hand side of the Attributes box. Notice the different Fields included and the Values that are given for the different volcanoes. Click on the names of various volcanoes listed and note that the values change. You may need to enlarge the Attributes dialog box by dragging the side or a corner of the box.

In what country is Stromboli located? Answer: _____

What type of volcano is Mt. Ararat? (Scroll down to find ARARAT, MT.)
Answer: _____

When did Vesuvius last erupt? Answer: _____

After you have answered the questions above, close the Attributes dialog box by clicking the X in the upper right-hand corner of the window. Click outside the map area to change yellow dots to red stars. Now you will select volcanoes on different tectonic plates and find out how many occur on a plate. You will use the Select Features tool again, but this time you will choose the Polygon shape. Before doing so use the Zoom In tool to enlarge the Caribbean plate, for example. Click on the Zoom In tool, and then drag a square to enclose the Caribbean plate, to enlarge it as shown here:

Zoom In

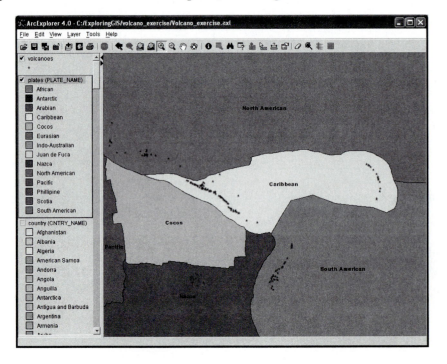

Next click in the Select Feature button on the toolbar and choose the Polygon. With the Caribbean plate enlarged, use the mouse and click around the plate boundary until you reach your starting point and double click; this will close the Polygon and highlight the volcanoes on the plate. Look up the Attributes by opening the Attributes window. After you have finished looking at one plate, use the Zoom to Full Extent button to see the entire map again. To deselect the highlighted volcanoes, double left-click outside the area of the map with the highlighted volcanoes.

Attributes

Zoom to
Full Extent

How many volcanoes occur on the Caribbean plate? Answer: _____

How many volcanoes occur on the South American plate? Answer: _____

How many volcanoes occur on the African plate? Answer: _____

When you have finished answering the questions above, Zoom to Full Extent.

Attributes of the Plate Boundaries Dataset
Notice that volcanoes occur within and at the margins of tectonic plates. However, they tend to form as lineations along plate boundaries. There are three types of major plate boundaries: divergent, convergent, and transverse. At divergent boundaries plates move away from one another. At convergent boundaries plates move towards each other. Transverse boundaries occur where two plates slide past one another; the boundary itself is a strike-slip fault. Volcanism occurs along both divergent and many convergent plate boundaries; however, at convergent boundaries, volcanoes occur parallel to but not directly on the plate boundary.

The Nazca and South American plates meet along a convergent boundary. The Nazca plate (moving relatively east) is subducting beneath the South American plate (moving relatively west). A chain of volcanoes occurs on the South American—the upper—plate. Using your map window, identify three other major convergent plate boundaries and circle them on the map below.

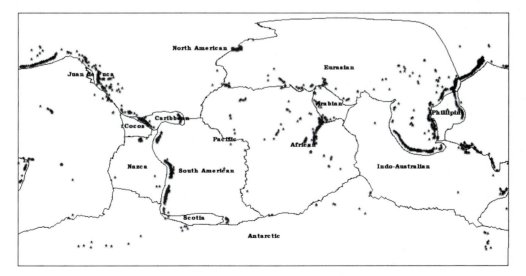

A tectonic plate is typically composed of both oceanic crust and continental crust. We can open the layer that shows the **country** boundaries to illustrate this fact.

First, remove the colors from the **plates** layer so that it is transparent. Open up the Layer Properties dialog for the **plates** layer. Choose One Symbol and Transparent fill for the Style. This will make the plates look white.

Click on the **country** layer. In the window that opens, you can see that most tectonic plates contain some continental crust.

Of the fourteen plates, five have no continental material on them. By clicking the **country** layer on and off and using the Zoom In and Zoom to Full Extent tools, determine which plates contain only oceanic crust and list them below.

Plates containing only oceanic crust:

      Answers: _____

_____

_____

_____

_____

Turn off the **country** layer for now. On the African continent there is a linear chain of volcanoes. These occur within the African plate rather than at the plate boundary. These volcanoes mark the location of the East African rift zone or spreading center. Make the **volcanoes** layer active and use the Select Features tool to highlight the volcanoes in this rift zone. Choose the Rectangle from the Select Features drop-down menu and draw a rectangle around the linear chain of volcanoes. Use the Zoom In tool to zoom into the area. Turn the **country** layer back on.

Select Features

Zoom In

What countries make up this rift zone? (Hint: Making sure that the country layer is active, either turn on the CNTRY_NAME under labels in the Layer Properties dialog box or use the Identify tool from the toolbar to help you.)

❶

Identify

Answer: _____

Volcanic Hazards and Populations
Turn off the **country** layer. If your **plates** layer is colored, turn off the colors. If the East African rift zone volcanoes are still yellow, click on the Clear All Selection

button on the toolbar. Turn on the **cities_1M** layer. This layer contains all the large cities of the world; each of the cites has one million or more residents.

To find out how many active volcanoes are near certain major cities, we will use the Query Builder tool. First make the **cities_1M** layer active. Then click on the Query Builder button. The Query Builder dialog box will open; the space circled in the figure below shows where you write the query. To write the query, double click on the first field, CITY_NAME, under Select a field. A dialog box will open that informs you that there are more than 100 volcanoes in this field and asks if you want to display all the values. Choose Yes. Single click on the equal sign. Next, single click on the city Jakarta (you will need to scroll down to find it), located under Values. Alternatively, you may also type it in as shown in the circle below.

Query
Builder

Your query should look like this:

Finally, click the Execute button.

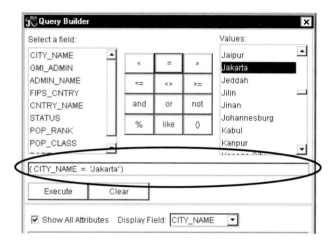

Close the Query Builder box by clicking the X in the upper right corner, or move the box to the side. Examine the map near the southern boundary of the Eurasian plate adjacent to the Indo-Australian plate. There you will see that one of the cities turned from blue to yellow. Jakarta is now selected. Use the Zoom In tool and drag a square around Jakarta to zoom into it.

With Jakarta selected, you can draw a buffer around the city to see how many volcanoes exist nearby. Click on the Buffer tool. Make a buffer distance of 100 miles around Jakarta. Check the box for selecting features from the volcanoes layer, since we want to know how many volcanoes are near the city. Use the figure below as a guide. Click OK.

Buffer

A red, semitransparent circle appears around Jakarta and the volcanoes within the buffer distance that you specified are now highlighted. Rather than count with your eyes the number of volcanoes highlighted within the buffer, use the Attributes tool. Since you are inquiring about volcanoes, activate the **volcanoes** layer and then click the Attributes tool button. As you can see, there are 15 active or recently active volcanoes within 100 miles of Jakarta, Indonesia. To execute additional queries, press the Clear All Selection button on the toolbar, choose the rectangle under Select Features, Zoom to Full Extent, and activate the **cities_1M** layer.

How many volcanoes are within 100 miles of Tokyo, Japan?
Answer: _____

Mexico City, Mexico? Answer: _____

Bangkok, Thailand? Answer: _____

Nairobi, Kenya? Answer: _____

Types and Shapes of Volcanoes

For this section of the exercise, you will examine the shapes of different types of active volcanoes by viewing digital elevation models (DEMs). Keep the **volcanoes** and the **cities_1M** layers turned on. Make sure that the volcanoes are denoted by red stars as these symbols will help you distinguish them from cities. Also, keep the **plates** and **country** layers turned on but make them transparent.

Add the DEMs to your project by using the Add Layers button on the toolbar. In the Catalog window under Data Sources, navigate to the folder where the project data are stored. If set-up properly in the first chapter, you should find the data in the C:\ExploringGIS\volcano_exercise folder. As shown below, change the file type to Image Files and choose all six files that have a suffix '.tif' and click the Add Layers button in the Catalog window. Do not add the item called '*ImageDirectory.' When done adding images, click the upper red X to close the Catalog window.

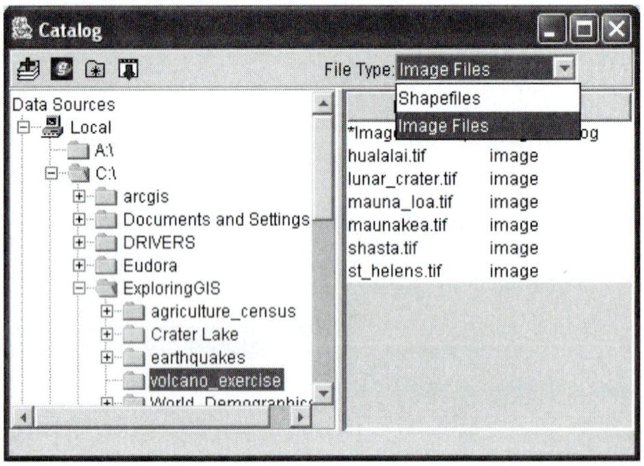

Add
Layers

Added to the top of your legend should be the files **hualalai.tif, lunar_crater.tif, mauna_loa.tif, maunakea.tif, shasta.tif,** and **st_helens.tif.** These images will be used for the last part of the exercise, and you should drag each one down to the bottom of your legend. In other words, the layers **volcanoes, plates, country,** and **cities_1M** should all be above the .tif images.

Check the box next to **st_helens.tif** to turn it on. Single click on this layer to make it active. Click the Zoom to Active Layer button. You should see a top-down (plan) image of a stratovolcano with a crater at the center. This is Mt. St. Helens in the state of Washington that erupted in 1980. You can see from the DEM that the eruption breached the north rim of the crater.

Zoom to
Active Layer

With the cursor centered for each click atop Mt. St. Helens, zoom out from Mt. St. Helens until you see two blue dots indicating cities. Note that the DEM will become miniscule. Two cities are approximately equidistant from Mt. St. Helens. Make the **cities_1M** layer active and use the Identify tool to find out their names. Use the Measure tool to measure in either kilometers or miles the distance from the volcano to both cites. To do this, click on one city and, while holding down the mouse key, drag a line to the volcano and then click again. In the box in the upper left-hand corner, read the measured distance. Clear the measure tool by choosing Clear Measure Totals in the drop-down menu of the Measure tool. Repeat process for the second city. Record below the names of the cities and their distances from Mt. St. Helens.

Zoom Out

Identify

Measure

Answer: City name _____ Distance: _____

Answer: City name _____ Distance: _____

Now activate, turn on, and zoom to **shasta.tif.** This is another stratovolcano—Mt. Shasta in California. The lobes on the north and northeast flanks of the volcano are lava flows.

Using the procedure you used for Mt. St. Helens, determine the identity of and distance to the nearest large city.

Answer: City name _____ Distance: _____

Now look at a different type of volcano. Turn on, activate, and zoom to **lunar_crater.tif.** This DEM has a pockmarked appearance and shows a volcanic field in Nevada that contains cinder cones. Now click on and make active the **volcanoes** layer and use the Identify tool to determine the latitude and longitude of the cinder cone called Lunar Crater, which you should see near the middle of the DEM.

Identify

Answer: Latitude _____ Longitude _____

What large city is closest to Lunar Crater and how far away is it?

Answer: City name _____ Distance: _____

The Hawaiian Islands offer good examples of shield volcanoes. Compared to stratovolcanoes and cinder cones, shield volcanoes have gently sloping flanks.

One at a time, activate, turn on, and zoom to **mauna_kea.tif, hualalai.tif** and **mauna_loa.tif.** Notice that on each of these shield volcanoes, you can see ropy structures approximately concentric with the top of each volcano. These are basaltic lava flows that are responsible for the gentle slope of shield volcanoes.

Use the Measure tool to measure in feet the length and width of the large crater at the top of Mauna Loa. Is this crater larger or smaller than the crater at the top of Mt. St. Helens?
Answer: Length _____ Width _____

Answer: <u>larger/smaller</u> (circle one)

With all three Hawaiian volcanoes and the country layer turned on, make active one of the Hawaiian volcanoes of your choice and zoom to that DEM. Zoom out from this until you can see all three of the Hawaiian volcano DEMs. If you've zoomed out enough, the outline of the Big Island of Hawaii will be visible. Note that one volcano occurs offshore and to the southeast of the Big Island. Make the **volcanoes** layer active and click on the Identify tool to find out the name and type of this offshore volcano.
Answer: Volcano name _____ Type of volcano _____

Identify

Close the project using the Close Project button or by exiting ArcExplorer. Do not save the changes.

Close
Project

What Have You Learned?
In this exercise you have studied the spatial distribution of volcanoes at the surface of the earth. In the process, you have learned about the composition of the earth's crust, the correspondence between volcanoes and plate tectonic boundaries, the shapes of different types of volcanoes. The exercise also gave you the opportunity to use a multitude of GIS tools including: Layer Properties, Select Features, Attributes, Zoom In, Zoom to Full Extent, Query Builder, Buffer, Measure, and Identify.

Exercise 6: Zebra Mussels: Mapping the Spread of an Invasive Species

Invasive species are one of the most persistent threats to biodiversity worldwide. An invasive species is one relocated, generally by human activities, to a new location where it has no natural predators. In general, this lack of predators allows invasive species to outcompete native or resident species, so that they spread rapidly.

Note that many relocated species do not become invasive or "weedy" species. But many species have advantageous traits, such as an ability to tolerate a range of ecological conditions or efficient methods of dispersing eggs, young, or seeds.

In this exercise, you'll map the spread of zebra mussels (*Dreissena polymorpha*), one of North America's most notorious aquatic invaders. These small mussels, usually less than 2 cm across, originated in Europe and were introduced to American waters in the ballast water of oceangoing ships. They are a plague to infrastructure, such as outflow pipes, grates, docks, and boats, and they also threaten aquatic ecosystems by filtering photosynthetic plankton from the water, aggressively colonizing surfaces underwater, and competing with native mussel species. As you make your maps, try to identify some of the shipping ports where the mussels were first introduced. As you work on this exercise, you'll also learn to use the Query Builder tool and export a table of data. Finally, if you have a spreadsheet program such as Excel, you can use it to graph the data table that you exported.

Zebra mussels were transported from Russia to North America in 1988. Since then they have spread steadily and caused billions of dollars of damage by colonizing water-intake pipes and restricting water flow. In this exercise you will examine confirmed zebra mussel sightings in the United States from 1988 to 2001. The source of the data is the U.S. Geological Survey (USGS), and data were acquired from the National Atlas of the United States (www.nationalatlas.gov).

Getting Started
To begin this project, copy the **zebra_mussels** folder from your CD to the ArcExplorer folder on your hard drive. This folder contains data showing the expanding distribution of this invasive species through freshwater systems of eastern North America.

Open ArcExplorer. From ArcExplorer, open the project called **zebra_exercise.axl:** go to File menu > Open Project (or Control O), then locate and select the map project in the ExploringGIS\Zebra_Mussel folder.

This project should open to display points where zebra mussels have been found, as well as county and province boundaries, and streams.

Save Your Own Copy of the Project
It's always a good practice to save your work early and often. Try saving a personal copy of your project. Go to the File menu, then Save As. Save your project in the C:\ExploringGIS\Zebra_Mussel folder, and give it your name, for example, "Jane zebra

mussel.axl." The ".axl" extension on the file name tells ArcExplorer that this is a readable project file.

Which Rivers Have Had Zebra Mussels?

Before you begin, make sure that your map is zoomed to the extent of the **zebra_mussels** layer. To see which streams have zebra mussels, you can use the MapTips button. In this case, you might want to display *river names* and the *dates* of confirmed zebra mussel colonization. Click on the MapTips button in your toolbar. Turn on river names by selecting the **rivers** layer on the left and the name field on the right. Then press the Set MapTips button. Before leaving the MapTips window, you can turn on the dates of confirmed zebra mussel sightings by selecting the zebra_mussels layer on the left and the Year field on the right.

Zoom to
Active Layer

MapTips

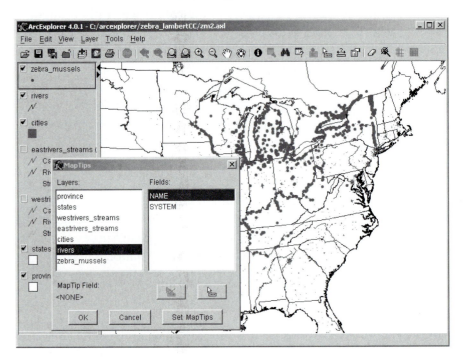

Now that you have set the MapTips, names and dates will pop up as you move your cursor over streams and zebra mussel sightings.

List three streams that contain zebra mussels and the year of the most recent sighting. (Remember, you may have to zoom in quite a bit to accomplish this.)

Streams   **Year**

_____   _____

_____   _____

_____   _____

Where Did They First Appear?

It is clear that zebra mussels have spread through many parts of the United States. However, their origin is less clear. In order to figure out where zebra mussels first appeared, it is necessary to build a query using the Query Builder.

Query
Builder

Select the **zebra_mussels** layer by clicking the layer name on the legend. You can tell it is selected when a black rectangle appears around it. Then click the Query Builder button in the toolbar, and select Year as the field. Since you would like to determine where zebra mussels first appeared, click on "=" and then on "1988." The resulting query equation should look like this:

(YEAR = 1988)

Next, press Execute, and the query builder will list all 1988 sightings. The same points will be highlighted on the map as well. Zoom to the active layer, so that you can see all the highlighted sites easily.

Zoom to
Active Layer

Where did zebra mussels first appear? Answers: _____

If these mollusks invaded from Russia, how did they get to Ohio and Michigan? What waterways might they have travelled?
Answers: _____

_____

Mapping the Dates of Mussel Appearance

As you know, the problem with zebra mussels is that they spread. In this portion of the exercise we will use Graduated Symbols to map the spread of this invasive species.

Double click on the **zebra_mussels** layer, to bring up the Symbols window. Set the symbol style to Graduated Symbols, then set the Field to Year. Now click Apply to see your changes without closing the Symbols window.

A range of colors should be applied to the year ranges. By selecting especially visible colors, you can make the range of years much easier to see. If you have a white background of states and provinces, a range from red to dark blue might be easiest to see, or perhaps blue to yellow. Try setting different Start and End colors until you have a range with clearly visible early and late dates.

As you set the colors, also notice that you can set the point size. By default, points are smaller for smaller values in the range, but this makes early dates (which are among the most interesting for us) hard to see. Set both ends of the range to the same size.

The following screenshot shows the ArcExplorer 4.0.1 application with the zebra_mussels properties dialog and a map of the eastern United States.

Set colors → Set dot size

Once you have chosen colors that are easily visible, edit the label of the first range. Select the label, and overtype "Less than 1991" to "1988 – 1991." This will make the legend more informative.

What port cities were first affected? To find out, turn on MapTips again, this time selecting **cities** as a layer and Name as the field to display. (Make sure MapTips are still on for zebra mussel dates and rivers.)

MapTips

List three of the first invaded cities:

Answers: _____

Answers: _____

Answers: _____

Now find some of the most recent invasions. Many recent colonies may be on smaller streams and tributaries of major rivers. To see these, turn on the layer **easternrivers_streams,** which shows small streams and canals. Up to now, you've had this layer turned off because it contains thousands of lines, and therefore it can be slow to draw. Set MapTips to Name for this layer. Then identify at least three streams or rivers invaded in 2000 or 2001.

MapTips

Recently invaded streams or rivers:

Answers: _____

Answers: _____

Answers: _____

Note that there are invasions more recent than 2001. However, the dataset was collected by the U.S. Geological Survey in 2001.

How Has the Invasion Changed over Time?

Next you can examine the growth of zebra mussels over time. It would be tedious to do this using the Query Builder for each year. Instead, you can export the data to a table in Excel.

If the Query Builder is closed, press the Query Builder button to reopen it. If it is still open, click the "Clear" button to remove the old data. This time you want to examine the data for all of the years, so enter (Year >= 1988). When you press Execute this time, the table will show each confirmed zebra mussel sighting for every year greater than and including 1988.

Query Builder

Clear All Selection

Press Execute. When you are working with a dataset containing thousands of points, you may see a message like that shown to the right.

Loading Query Results

? There are more than 100 records selected. Do you want to see all records?

Yes   No

In this case, you DO want to see all the records, so click Yes. (If you click No, you'll only export a small part of the table.)

All the zebra mussel points are now selected, and they will be highlighted on the map as well as displayed in the Query window. The lower portion of the Query Builder box shows information about each site, including the data and location by state, county, and stream. Note: This information will not be displayed if the Show All Attributes box is not selected.

Export a Table

Now you need to export all of this information into an Excel table. In the lower right-hand corner of the Query Builder box is an image of a disk. Click on the disk, and save the file as a text: first make sure you are saving the file in the folder you are working in (ExploringGIS\Zebra_Mussels\, then in the File name field, type **mussels.txt.** The ".txt" indicates that this is a plain text file. You could look at it in a word processing program if you wanted to, although it would be hard to read.

Now open Excel. (Other database or spreadsheet programs will work as well. The directions that follow are for Excel, which is widely used on academic computers.)

Under the File menu select Open, and find the mussels.txt file that you just created. Cells in the table are separated by commas, so you need to specify "delimited." Then click Next and specify "comma." Then click Finish. A table will be produced with several columns that replicate the information from the Query Builder.

Excel Import window, indicating that the text file is Delimited:

Excel Import window, indicating that the file is delimited by Commas:

When Did the Invasion Peak?

Now you can figure out when the invasion of zebra mussels peaked. The easiest way to analyze this data is to make and examine a histogram. There are several steps to this procedure.

1. Create a list of "bins," which are the classes by which your data will be grouped. Your bins will be years, since you want to see the abundance of new sightings by year. You have data for the years 1988 through 2000, so find an empty column (such as column M, at the right side of the spreadsheet) and fill in 13 cells with these dates.

2. Under the Tools menu select Data Analysis.
    Note: If you do not see Data Analysis listed under the Tools menu, you can add it by selecting Add-Ins under the Tools menu. Check the box for Analysis ToolPak and press OK. This will add the Data Analysis tool to your Tools menu.

In the Data Analysis box, select Histogram. The Input Range should be the extremely long list of years that was produced from the initial Query Builder. It should be in column B. Select this column by clicking the top cell (B). Move the cursor to the Bin Range box. The Bin Range should be the column of 13 date cells that you just made. Select it by clicking the top cell (M, if that is the column you used). Under Output options, select New Worksheet Ply, then press OK. Usually this step takes awhile to process, so do not be alarmed if you have to wait.

Eventually a new sheet will be produced that has two columns for Bin and Frequency that represent the year and number of sightings respectively.

3. Now you can graph and inspect these data. Under the Insert menu, choose Chart and select a Chart type. Use the Clustered Column Chart. Keep your series in columns rather than rows as it is generally easier to read. Add a title to your chart, such as "Rates of Zebra Mussel Invasion." Specify the category of the X-axis (year) and the value of the Y-axis (frequency). Finally, create your chart on a new sheet and analyze your findings.

Which year had the most sites? Answer: _____

Which year had the least sites? Answer: _____

Approximately how many sightings of zebra mussels were made in 1993?
Answer: _____

Does the chart look how you imagined it would? Why or why not?
Answer:_____

_____

Note: The data in use were published in 2000 so the small number of observations from that year isn't necessarily realistic. Using the surrounding years as a guide, what do you think would be a more realistic frequency level for 2000?
Answer: _____

Are Zebra Mussels a Problem in Your Area?
If you live in the eastern United States or Canada, you might be interested in learning whether or not zebra mussels are a problem in your area. You can use the Find tool on the toolbar to find out. (If you live in a western state or province, choose an eastern location to inspect for this part of the exercise.)

Find

First turn on the **counties** layer. For easier visibility and faster display, you may want to turn off the **eastrivers_streams** layer. Then click on the Find tool.

In the Find window, select **counties** as the layer to search in, then enter your county's name in the Value space. Remember that this field is case sensitive so use correct capitalization. Then press Find.

Sometimes several counties with the same name will appear. If this happens, simply click on each one and pan to its location. Note that the Pan tool centers the selected county in the window. The Zoom In tool zooms in on the selected county. When you see the location change to your state, zoom in. Then exit the Find box.

If you're not sure you have the right county, you can check the name and state of your location by using the Identify tool on the top toolbar. You must make sure that the **county boundaries** layer is active on the left legend, then click on the location.

Identify

Once you have the correct county, zoom out until you can see the nearest zebra mussel site.

Zoom Out

Are there any sites in your county? _____

If not, look for the nearest zebra mussel site. Use the Measure tool on the top toolbar to draw a line from your county to the site. (Remember, to use this tool, you must first check its units. In this case, your data are measured in meters. Go to the View menu > Scale Bar Properties > Map Units > Meters.)

Measure

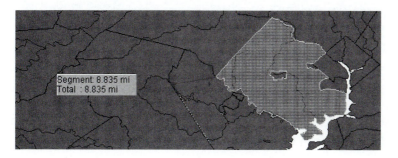

How close are zebra mussels to your location, in km? _____

Invasion of Chicago

For one last step in this exercise, take a look at zebra mussel invasion dates in Chicago, centering on Cook County, Illinois. Do a search like the one you just did: use the Find tool again, select **counties** as the layer to search, and enter Cook in the Value space.

Find

This time you will find that there are several Cook counties. Without leaving the Find window, select the different Cook Counties and click Pan. This will center each in the map window. Keep doing this until you find the one in Illinois.

Pan

Once you have found Cook County, Illinois, you can see when the Chicago area
became most badly affected by zebra mussels. First you should reset the Symbols of the
zebra mussels to individual dates: double click on the **zebra_mussels** layer to open its
Symbol properties. Set the Symbol to Graduated Symbols, using the Year field. Set the
color range from dark blue (minimum) to red (maximum). Set the number of classes to
15. Set the size of all symbols to 5 pt.

For reference, you may choose to turn on the **cities** layer, to make Chicago visible, as
well as Cook County.

With 15 colors, you won't be able to distinguish visually one year from another, but
you can select the years. Starting at the top of the list, click on the symbol in the table
of contents. This will highlight all the points with that value.

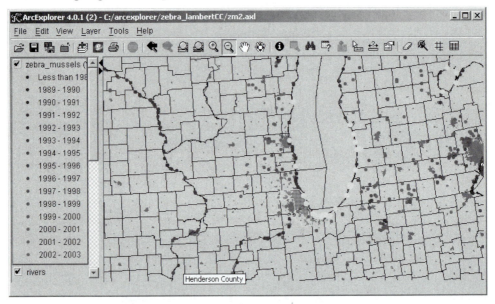

In what year do zebra mussels first appear in Cook County, Illinois?

_____

When did the mussels become abundant in the Chicago area?

_____

When did zebra mussels spread up the Chicago River, which runs southwest from the city?

_____

Finish Your Project
Find the year when there were the most new invasions in the Chicago area. Print your
map (press the Print tool, and to be safe, specify that you are printing only 1 page).
Note that highlighted dots will show up better on a color printer, but they can be seen in

Print

a black and white print. Write the date at the top of your map and turn it in to your instructor if requested.

Save your work under the name you gave it at the beginning of this exercise. Close ArcExplorer.

What Have You Learned?
You have practiced a number of important skills in this exercise. You have inspected your data using MapTips, and you have assigned graduated colors to the dates of zebra mussel invasions. You have performed a query on the attributes of the zebra mussel data, exported your query results as a table, imported the table to Excel, and created a histogram to show the rate of mussel expansion over time. You have used the Find tool to search for features in your map (in this case, finding counties by name). And you have selected attribute values in your table of contents to display dates in a local area.

Exercise 7: Global Air Quality and the Kyoto Protocol

Many human activities result in the emission of "greenhouse gases," gases named for their ability to trap heat in the earth's atmosphere and cause warming more-or-less as in a greenhouse. Most scientists concur that these gases have contributed to global climate change. Recent climate change has occurred at an alarming rate compared with climate changes observed in the geologic record, and three of the warmest years on record occurred in the past 10 years. These changes were acknowledged at the 1992 United Nations Framework Convention on Climate Change (UNFCCC), in which conveners considered the effects of gas emissions on the climate system. However, this international convention did not address particular gases or provide guidelines to reduce emissions of these substances.

As concern grew along with greenhouse gas emissions, negotiations began toward a binding agreement among countries, and the Kyoto Protocol to the UNFCCC was established in December of 1997 in Kyoto, Japan. At the meeting, participating nations agreed upon a treaty to work collectively to decrease global greenhouse gas emissions. According to the Kyoto Protocol, industrialized countries, termed "Annex I Parties" in the treaty, would reduce emissions of greenhouse gases by at least 5 percent below each country's 1990 emissions levels by 2012. Developing countries will be exempt from emissions limits. For the Protocol to become a binding international agreement, it must be signed by countries representing 55% of emissions.

The future of the Protocol is in doubt, as two of the largest greenhouse gas producers, the United States and Russia, appear unlikely to support it. Nonetheless, this accord has been an important step toward global efforts to reduce human impacts on our climate. To view the current status of ratification of the Kyoto Protocol, visit the website http://unfccc.int/resource/kpthermo.html.

In this exercise you will analyze four greenhouse gases addressed by the Kyoto Protocol—carbon dioxide (CO_2), methane (CH_4), nitrogen oxides (NO_x), and chlorofluorocarbons (CFCs). The table below shows the sources of each gas as well as the degree to which it contributes to global warming.[1] As is evident from the table, carbon dioxide contributes most significantly to global climate change.

Greenhouse Gas	Proportion of Contribution to Global Warming	Sources
Carbon dioxide (CO_2)	64%	Respiration, fossil fuel burning, land clearing, industrial processes
Methane (CH_4)	19%	Transportation, power plants, rice paddies and wetlands, gas drilling, landfills, ruminant animals
Chlorofluorocarbons (CFCs)	11%	Spray propellants, refrigeration compressors, foam blowing
Nitrogen oxides (NO_x)	6%	Fossil fuel burning, fertilizer, lightning, biomass burning, soil microbes

---

[1] Cunningham, W.P., M.A. Cunningham, and B.W. Saigo, *Environmental Science: A Global Concern*, 7th ed. (Boston: McGraw-Hill, 2003).

For additional information on climate change and greenhouse gases, please consult your course textbook or discuss the topics with your instructor.

Getting Started
Before starting this exercise, copy the directory **Air_quality_exercise** from your CD to the C:\ExploringGIS\folder on your computer.

Open ArcExplorer. (Start > Programs > ESRI > ArcExplorer) Click on the Add Layers icon and navigate to the Air_quality_exercise folder. Add the shapefiles **CFCs, NOx, CO2,** and **Methane** to the legend by double clicking on each one in the Catalog window. Remember that if you see an error message while opening the shapefiles, just press Continue. Once each shapefile has been added to the legend, close the Catalog window. Organize the layers in your legend so that **Methane** is at the top, **CO2** is second, **NOx** is third, and **CFCs** is on the bottom. Once again, go to Add Layers and add another **Methane** shapefile, which will show up in the legend as **Methane0**. Once **Methane0** has been added, close the Catalog window. **Methane** and **Methane0** are the same file; however, you now have a second copy with which to work. Click on the **Methane** layer to make it active and check the box on the layer to show the world map.

Add Layers

Add Layers

Greenhouse Gases
Since the Kyoto Protocol uses a country's emissions in 1990 as a guideline for reducing emissions in the present and future, the greenhouse gas data you will analyze will normally compare one year of emissions with those in 1990. Let's examine methane emissions.

Methane
In this part of the exercise, you will colorize your world map based on 1990 methane emissions. Right-click on the **Methane** layer to go to Layer Properties or click on Layer Properties on the toolbar. Draw features using Unique Symbols, from the drop-down list adjacent to Field for values, choose Total_1990 (it is near the bottom of the list of fields), and use the Pastels Color Scheme. That is, duplicate the Properties window shown below and click OK.

Layer
Properties

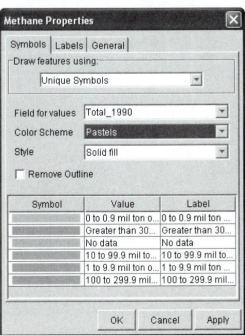

Your map will be updated with a color coding that shows total methane emissions for 1990. You will see in the legend six ranges from "0 to 0.9 mil ton of CH_4" to "Greater than 300 mil ton of CH_4." There is also a "No data" range. To make the map easier to read, let's change the color coding. Right-click on the **Methane** layer to go to Layer Properties or click on Layer Properties on the toolbar. Change the color for "No data" to white and for "Greater than 300…" to red. To do this, click once on the color next to the value you wish to change. In the Color Chooser window that opens, click on the color that you would like to use for the value and then click OK. Change both values and then click OK to view the updated map. By changing the color scheme in this way, the countries with greater methane emissions will be most visible.

Now repeat these steps for the second copy of the **Methane** shapefile in the legend, but this time set the Field for values to be Total_1995.

Click the topmost **Methane** layer to turn it on and off and compare visually, methane emissions in 1990 and 1995. What four countries appear to have been the largest methane emitters in 1990? Use the Identify tool to help you, but remember that to use it, a **Methane** layer must be active.

Identify

Answer: _____

                   _____

                   _____

                   _____

To determine specifically which countries emitted the most methane in 1990, use the Query Builder tool. As shown below, choose the field Total_1990, click on the "equals" sign and choose the value "Greater than 300…" You do not need to type in the formula shown in the circle. Click Execute.

Query
Builder

Query Builder

Select a field:

| ENERGY_95 |
| AGRICULT_9 |
| WASTE_95 |
| OTHER_95 |
| TOTAL_95 |
| Total_1990 |
| Total_1995 |
| Change |

<	=	>
<=	<>	>=
and	or	not
%	like	()

Values:

0 to 0.9 mil ton of CH4
1 to 9.9 mil ton of CH4
10 to 99.9 mil ton of CH4
100 to 299.9 mil ton of CH4
Greater than 300 mil ton of C
No data

(Total_1990 = 'Greater than 300 mil ton of CH')

| Execute | Clear |

☑ Show All Attributes   Display Field: CNTRY_NAME ▾

Four countries are listed under Query Results along with many columns of data. Resize the Query Builder window so that you can see more of the data. The first four columns reveal for 1990 the amount of methane emitted by energy, agricultural, and waste sectors as well as sources of methane characterized as "other." An additional column shows total methane emissions for 1990.

Use these data to calculate the percent of methane that each sector contributes per country. For example:

$$\% \text{ Methane from Energy\_90 }_{(country\ name)} = (ENERGY\_90/TOTAL\_90) \times 100$$

$$\% \text{ Methane from Energy\_90 }_{(Russia)} = (473.7/631.4) \times 100$$

Record your answers in the table below.

Country Name	% Methane from Energy 1990	% Methane from Agricultural 1990	% Methane from Waste 1990	% Methane from Other Sources 1990

Note that the Query Builder also lists data for 1995. Therefore, you can see which countries have reduced methane emissions and which have increased them over the 5-year span represented by the data. The last column is listed as Change. For each country, this column of data was calculated by subtracting the value of the TOTAL_90 field from the value of the TOTAL_95 field. A negative number indicates that the country reduced its methane emissions while a positive number shows that the country increased methane emissions from 1990 to 1995.

Click the Clear button and build a query that asks for a value greater than five in the Change field. What countries were selected?

Query Builder

Answer: _____   _____

                 _____   _____

                 _____   _____

                 _____   _____

Close the Query Builder window. Remove the two **Methane** layers from the legend by clicking on Remove Layer from the Layer drop-down menu in the toolbar.

Carbon Dioxide
Now you will scrutinize global carbon dioxide emissions by constructing a graph that shows those countries that have emitted the greatest amounts of carbon dioxide. Make the **CO2** layer active and click on the Query Builder tool.

Query Builder

The Query Builder window shows carbon dioxide data for 5 years. The dataset provides CO_2 emissions by gas fuel consumption in thousand metric tons of CO_2.

Build a query that retrieves the two greatest emitters in 1965. To do this, click once on the field Y_1965. All values for 1965 will appear in the right-hand side of the Query Builder window. Scroll down and find the last two (highest) values. Use the 'greater than or equals' symbol and the second highest value to determine the two countries with the highest CO_2 emissions in 1965. Write these names below.

Answer: _____

_____

Next, you will plot the data for these countries on a graph. These longitudinal data will help us see how countries have changed their emissions over time. If you have Excel or some other spreadsheet program, you can save the data from this window and use it in the graphing software package. If you do not have graphing software, write down the numbers shown in the Query Builder window.

Click the Save button in the lower right-hand corner of the Query Builder window (shown above). You will be prompted to save the file. Navigate to the working folder that you have been using for this exercise, give the file a useful name, like 1965, and give it a *.dbf extension. Do these steps for each of the years (1965, 1975, 1985, 1995, and 1998), to create five *.dbf files. If you notice that two countries with the greatest carbon dioxide emissions for a given year have been repeated from another year, you don't need to save that file a second time. Remember to click the Clear button for each new query. An entry of "-999" indicates that no data are available for that country for that year.

To create the carbon dioxide emissions graph, open Excel and go to File > Open file or Control O. Choose All File Types from the drop-down menu at the bottom of the Open window. Next, navigate to the folder that contains your saved queries. Open them one at a time. Each time you open a new file, you will be prompted with a Text Import Wizard. When requested, indicate that your files are delimited by commas. As you open them, your files should be arranged in columns just like the results of your queries. Notice, however, that the heading that indicates the year of the data is not present. With your Excel spreadsheet open, insert a line above the first row and type in the proper year at the top of each column of data. Copy and paste the name of each country and its row of data into one worksheet; each country and its respective data should appear only once in your spreadsheet; that is, do not duplicate rows.

Next, drag a box around the data in your Excel spreadsheet. Click on the Chart Wizard icon:

 Excel's Chart Wizard tool

For Chart type choose Line and use the default settings for each of the windows that follow. This will create a line graph of your data. What trends do you see from the four countries with the greatest carbon dioxide emissions over the past 30 or more years? Explain why some data are missing.

Answer: _____

_____

_____

Print out your line graph and turn it in with this exercise. Exit Excel.

Remove the **CO2** layer from your legend in ArcExplorer.

Nitrogen Oxides
Make the **NOx** layer active. This data set contains nitrogen oxide emissions for the years 1990 and 1995 as well as the change in amounts of emissions over the 5-year period. The units are in thousand metric tons of NO_x. In this part of the exercise you will map the change in nitrogen oxide emissions between 1990 and 1995. As with the carbon dioxide section of this exercise, a negative number indicates that a country has reduced its nitrogen oxide emissions while a positive number indicates an increase in nitrogen oxide emissions.

Layer
Properties

Right-click on the **NOx** layer and go to Layer Properties or choose Layer Properties from the toolbar. Under the Symbols tab, Draw features using Graduated Symbols and choose the NOx_change field. Keep all the default settings and click OK. Make sure that the NOx layer is visible by checking the box adjacent to it. You should now have a colored world map that shows countries according to the change in nitrogen oxide emissions between 1990 and 1995. Countries that have substantially reduced their emissions are shown in yellowish orange and yellow whereas countries that have increased their nitrogen oxide emissions between 1990 and 1995 are show in reddish orange and red. Use the Identify tool to determine the names of those countries that have increased their NO_x emissions between 1995 and 1990. Record your answer.

Identify

Answer: _____

Sometimes subtleties in data may be difficult to detect visually. Use the Query Builder tool to find the countries that have shown a marked increase in nitrogen oxide emissions from 1990 to 1995. Build the query to find out how many countries had an emissions change of more than 300 units. There will be more than 100 values when you choose NOx_change as the field and you will be prompted to answer whether you want to display the values; click Yes. According to the Query Builder, how many countries have increased their NO_x emissions by more than 300 units between 1990 and 1995 and what are their names?

Query
Builder

Answer: _____

Close the Query Builder window. Remove the **NOx** layer.

Chlorofluorocarbons
The **CFCs** shapefile contains data on the emissions of CFCs, or chlorofluorocarbons, given in units of Ozone Depleting Potential (ODP) tons for the years 1990 and 1995. In this portion of the exercise, you will construct two maps and visually compare CFC emissions during these two years.

Turn on and make active the **CFCs** layer. Using Layer Properties, change the world map to show Unique Symbols for the Total_1990 field (do not use the Y_1990 field). Use Minerals for the Color Scheme. And as shown below, change the color for "No data" to white, change the color for "0 ODP tons" to a light color, and change the color for "Greater than 10,000 ODP tons" to bright red.

Layer
Properties

Click on the color
under the word
"Symbol" to ⟶
change the color.

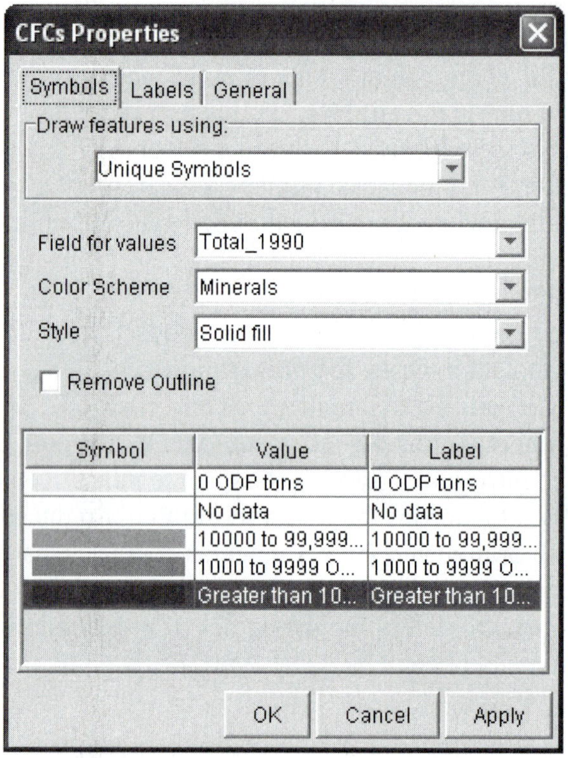

Now click OK to see your world map of global chlorofluorocarbon emissions for 1990. You may need to resize the legend to read the meaning of each color. If you lose part of the map when you resize the legend, click the Zoom to Full Extent button on the toolbar to view once again the entire world on your map.

Zoom to
Full Extent

Your map is easier to read as a result of the changes you have made. However, these changes also highlight a limitation of GIS. Where no data exist, it is not possible to make complete maps. Also, ideally, all countries supply accurate data on their emissions, so that emissions reported as "0" really mean no emissions of the gas of concern. Sometimes though, countries report "0" to mean no data rather than zero emissions and thus, unless the data are scrutinized carefully, a map may be misread.

To illustrate this point, Zoom In on the countries of Europe. Use the Zoom In tool to draw a box around most of the countries of Europe. Though these countries are highly industrialized, as you can see, few of them seem to have reported any chlorofluorocarbon emissions. In fact, they must have reported their CFC emissions as "0" to indicate that they have no data. Now zoom back out to the full extent of the map.

Zoom In

Zoom to
Full Extent

We will now compare the 1990 chlorofluorocarbon emissions map with a map prepared using data from 1995. To do this, open a new ArcExplorer window by going to File > New ArcExplorer or Control N (as shown below).

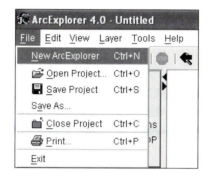

In your new ArcExplorer window (which should have a '(2)' in the title bar), add the **CFCs** layer by choosing Add Layers from the Layer drop-down menu and navigating to the folder that contains the CFCs shapefile. Double click on the shapefile to add it to the legend of your new ArcExplorer window. Click the box to turn on the **CFCs** layer.

Add
Layers

Create a map just as you did above by choosing Layer Properties from the toolbar. Once again Draw features using Unique Symbols but instead of using the Total_1990 for Field for values, use Total_1995. Do not use the field Y_1995 for this map. Change the map colors to match those used for the 1990 emissions map to make it easy to compare the 1990 and 1995 maps. By examining the two, you will be able to see how countries have changed their reported CFC emissions over the 5-year period we are examining.

Layer
Properties

In order to compare the two maps, resize each of the map windows by dragging a corner so that both windows will fit on your computer screen. Or, flip back and forth between the two maps.

Because you have paid close attention to the colors of your maps the two maps should look similar. How similar are they? Which map has more countries colored in bright red, 1990 or 1995?
Answer: _____

Exercise 8: Water Quality and Environmental Health

Getting Started

This exercise posits a hypothetical situation: you would like to purchase land that will provide your family with opportunities to fish and swim in a stream on your property. Additionally, you would like the land to afford some privacy. In order to find such a place, you need to locate land for sale that has a stream running through it and you want to confirm that the stream water is clean. The following exercise illustrates how one can locate land with particular characteristics and also assess surface water quality for local bodies of water. The data you will use might pertain to any location where streams flow through residential areas.

Before starting this exercise, copy the directory **Water_Quality_exercise** from your CD to the C:\ExploringGIS\folder on your computer. Open ArcExplorer (Start > Programs > ESRI > ArcExplorer). Click on the Add Layers icon and navigate to the Water_Quality_exercise folder. Add the shapefiles **Homes_For_Sale, river,** and **Streets** to your ArcExplorer legend by double clicking on each shapefile and then close the Catalog window. Remember that if you see an error message while opening the shapefiles, just press Continue. Click on each box to make these data visible in your map window. To modify the display of the data, right click on the shapefile in the legend that you would like to change and open the Layer Properties dialog box, or use the Layer Properties tool on the toolbar. For example, for the **river** shapefile, change the line color for the river to blue and the size to 2 as shown below.

Add Layers

Layer
Properties

river Properties

Symbols | Labels | General

Draw features using:

One Symbol

Style — Solid line

Color — ■ Blue

Size — 2

OK | Cancel | Apply

Similarly, adjust the **Streets** file to show gray lines with a line size of 1 and switch the **Homes_For_Sale** point data to display brown squares in size 8.

Your map window now displays a subset of a larger dataset of streets and rivers. These data were clipped—that is, extracted—from a larger dataset using another ESRI

Zoom to
Full Extent

software called ArcGIS. The fact that they were clipped accounts for the circular shape of the map. If the circular shape of the map is not obvious, use the Zoom to Full Extent tool.

Locating Homes for Sale
Let's assume that given your desire for a home near a stream, you have narrowed your search to a particular area and have found an appealing region in which homes for sale might be located near a stream.

First, you need to identify homes for sale that meet the criterion of being along a stream. To do this you will buffer the **river** file and then select homes from the **Homes_For_Sale** file.

Click once on the **river** file to make it active. Using the Select Features tool from the toolbar, choose the circle shape. Place the cursor in the middle of the map and drag a circle shape so that all of the **river** lines are selected. The circle that you draw will be red and the **river** lines will turn bright yellow once they have been selected. With the **river** file selected, you can buffer the river to a distance of 450 feet. That is, assuming you desire a home no farther than 450 feet from the river, you will create a buffer of 450 feet around the river and see what homes fall within that buffer zone. The software will select the homes for you.

Select Features

With the **river** layer active, choose the Buffer tool from the toolbar. As illustrated below, in the Buffer window that opens, type in 450 for the Buffer Distance, choose "Feet" for the Buffer Units from the drop-down menu, check the box Use buffer to select features from this layer, and elect the **Homes_For_Sale** layer. Click OK.

Buffer

Buffer			
Buffer Distance:	450		
Buffer Units:	Feet		
☑ Use buffer to select features from this layer.			
Homes_For_Sale			
	Clear Buffer		
OK		Close	Apply

Notice that a red buffer line has been drawn around the river and that some homes have turned from brown squares to bright yellow ones. Since they are difficult to see, have the software locate them for you. To do so, click once on the **Homes_For_Sale** layer to make it active and choose the Attributes tool from the toolbar. The Attributes window lists twelve properties for sale within 450 feet of the river. Write down the addresses of these properties in column A of Table 1 at the end of this exercise.

Attributes

Though the selected homes lie within 450 feet of the river, some may have no river access. That is, a street may run between the house and the river. Use the Identify and Zoom In tools on the toolbar to identify each location by address and determine if the home has direct access to the river. For homes with river access, enter "yes" in column B of Table 1 and for those with no access enter "no."

If the Attributes window is still open, close it and use the Clear All Selection tool to clear the buffer and its accompanying selections.

Evaluating Land Use

The houses you have identified will be of interest as potential homes if they are on land zoned for residential use or categorized as forested land or cropland. Properties zoned for mixed commercial use are not ones you would like to purchase. Therefore, in this portion of the exercise, you will determine the local land use for each of the properties for sale.

Using the Add Layers tool, add to your map the **LandUse** shapefile from the Water_Quality_exercise folder. This layer contains categories of land uses that you can view easily using multiple colors. Right click on the **LandUse** layer to open the Layer Properties window or choose Layer Properties from the toolbar. As shown below, click the Symbols tab. Draw features using Unique Symbols, choose Code_name for the Field for values, select the Random Color Scheme, and use Solid fill for Style. The software colorizes the values randomly. Change the colors by double clicking on the color symbol so that Lakes and Reservoirs are blue and Forest is green. Also, check the Remove Outline box to blend together the land use polygons of similar value and thus make the **LandUse** layer most readable.

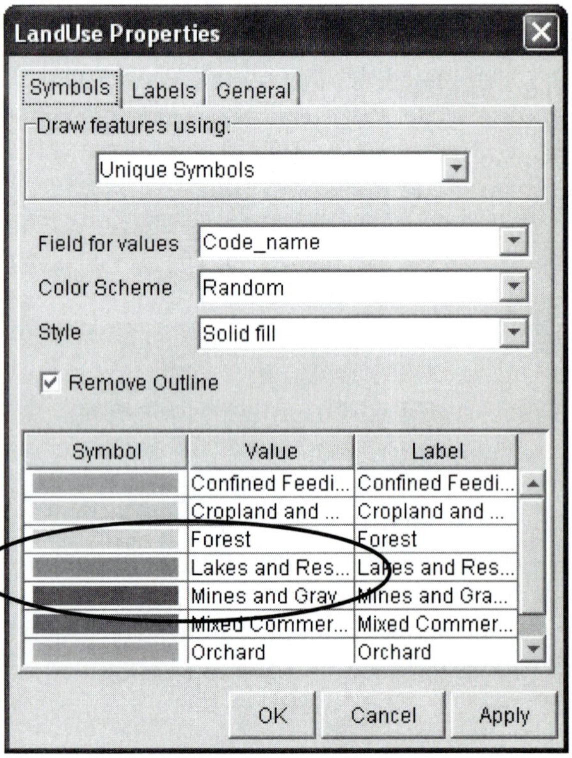

Change Lakes and Reservoirs to blue, Forest to green.

Identify

Zoom In

Clear All Selection

Add Layers

Layer Properties

With **LandUse** as the topmost layer, you can determine the primary land uses in this area by observing which colors dominate the **LandUse** layer. If it is difficult to read the types of land uses from the legend, widen the legend area as follows: place the cursor between the legend and the map area, look for the horizontal line with arrows at both ends and drag the legend to expand it as shown below.

To make the legend area larger drag from the area between the legend and the map region (circled).

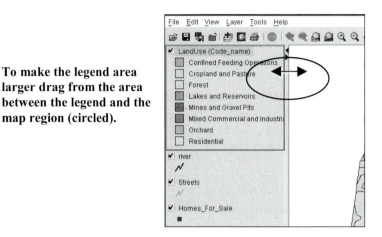

What are the two predominant land uses in this area?
Answer: _____

Next, you will determine the type of land use for each of the properties you have listed in Table 1. Drag the **LandUse** layer to the bottom and turn on all layers. Once again buffer the river to a distance of 450 feet as you did in a previous section of this exercise. Use the MapTips and Identify tools, to determine the address and land use type for each selected house denoted in yellow. Note that when you choose MapTips from the toolbar, a dialog box will open. Choose the **Homes_For_Sale** layer and the ADDRESS field, click Set MapTips, and OK. You will need to toggle between **Homes_For_Sale** and **LandUse** in the legend to read the address and the land use type in the Identify Results dialog box. Write the land use code name for each of the properties under column C in Table 1.

MapTips

Identify

When you are finished checking land use, turn off or remove the **LandUse** layer and click Clear All Selection.

Clear All Selection

Assessing Water Quality
In this portion of the exercise you will analyze water quality in the river and tributaries that flow through the region in which you may purchase a home.

Water quality samples were collected at twelve sampling stations along the river and the data are included in the GIS software. Water quality parameters that were analyzed include temperature, stream flow, specific conductance, dissolved oxygen, and pH. These are standard water quality indicator parameters that reveal the suitability of water for purposes such as swimming and fishing as well as drinking. Ideally, an interested person would monitor water quality data for at least a year to evaluate seasonal effects. In this instance, we will examine data for only one sampling event. Thus the data reflect water quality at one moment rather than over an extended period of time.

Parameters such as temperature and stream flow rates for the twelve sampling sites in this exercise appear to be uniform across each of the water collection stations. Therefore, these parameters provide little insight into variations in water quality along the properties you might purchase. Rather, variations in specific conductance, dissolved oxygen, and pH may reveal useful information about water quality on the properties.

Specific conductance measures the flow of electricity through water over a set distance of water sampled, in this case 25 centimeters. High specific conductance indicates high salt content (salinity) of water. The presence of dissolved salts in freshwater streams negatively impacts freshwater aquatic life.

Oxygen is essential for aerobic aquatic life. Therefore, for a stream to support such life, it must contain a relatively high amount of *dissolved oxygen.* Dissolved oxygen levels of 6 milligrams per liter or higher indicate good quality water.

Water acidity or alkalinity is measured by *pH*—a scale that runs from greater than 0 to less than 14. pH values less than 7 indicate acidic water and those greater than 7 are basic. Extreme values of pH limit the ability of aquatic plants and animals to thrive.

In this section, you will make maps that display the specific conductance, dissolved oxygen, and pH in order to assess water quality. The following table shows the units of measurement and ArcExplorer Field Names for each of the water quality parameters:

<table>
<tr><td colspan="2">Water Quality Parameters Units of Measurement
and ArcExplorer Field Names in Parentheses</td></tr>
<tr><td>Temperature (TEMP)</td><td>degrees Celsius</td></tr>
<tr><td>Stream flow (STREAMFLOW)</td><td>cubic feet per second</td></tr>
<tr><td>Specific Conductance (CONDUC)</td><td>microsiemens per square centimeter</td></tr>
<tr><td>Dissolved Oxygen (DIS_OXY)</td><td>milligrams per liter</td></tr>
<tr><td>pH (pH)</td><td>standard units</td></tr>
</table>

Click the Add Layers button from the toolbar and choose the sample_points shapefile from the Catalog window. Double click on the sample_points shapefile three times in order to add three **sample_points** layers to the map legend and then close the Catalog window.

Add
Layers

The three **sample_points** layers will appear in the legend as shown below.

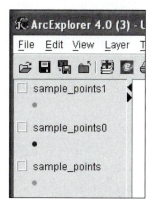

Note that two of the **sample_points** layers will have the numbers "0" and "1" after them and the third will have no numerical designation.

Let's begin by mapping specific conductance. Right click on one of the **sample_points** layers and choose Layer Properties. Guided by the figure below, Draw features using Graduated Symbols, select the Field CONDUC, choose 3 Classes and Circle for Style, use start and end default colors (yellow and red), and change the size ranges to start at 5 and end at 15. Click OK.

Layer
Properties

sample_points Properties

Symbols | Labels | General

Draw features using:

Graduated Symbols

Field   CONDUC
Classes 3   Style Circle

Color   Size

Start ☐ Yellow   Start 5
End ■ Red   End 15

Symbol	Range	Label
	Less than 500.0	Less than 500.0
	500.0 - 640.0	500.0 - 640.0
	640.0 - 780.01	640.0 - 780.01

OK   Cancel   Apply

Make sure the **sample_points** layer is active and checked so that you can inspect the data on your map.

Also, identify the sample station numbers as follows. Once again choose Layer Properties, go to the Labels tab and label features by Station.

Layer Properties

On the Water Quality Worksheet, Table 2, at the end of this exercise, mark an X in the Specific Conductance column (B) for the stations that have the highest values of conductivity. If you followed the steps above and used three classes, the high values for specific conductance will appear on the map as large orange or red circles.

Now follow the same steps for dissolved oxygen (DIS_OXY) and pH. Use three graduated symbols as you did for specific conductance but choose different styles for these parameters such as triangles and stars, respectively. When you evaluate each of these parameters, you may want to turn off the other two water quality maps. On Table 2, mark with an X those stations that have low levels of dissolved oxygen and pH values higher than 7.5.

Examine your data as reported in Table 2. Water monitoring stations with two or more Xs in columns B, C, and D indicate lower water quality than is preferable for recreational purposes such as swimming and fishing. In column E of Table 2, write the word "poor" for stations that have two or more Xs in columns B, C, and D. Note that the stream in your area flows from north to south. For example, water flows from Stations 9, 12, and 8 past the other stations and ultimately down to Station 1.

Now that you have assessed water quality at stations in the watershed, use these data to narrow your choices of possible home purchases. Houses adjacent to or downstream from sampling stations with low water quality should be eliminated from the list of desirable properties.

MapTips

With the **Homes_For_Sale** layer active, choose the MapTips tool. Using **Homes_For_Sale** for the layer and ADDRESS for the field, click the Set MapTips button and click OK. Now use the Identify tool to locate addresses of houses downstream or adjacent to water stations with poor water quality. Write "poor" in the Water Quality Evaluation column (D) on Table 1 for these addresses.

Identify

Turn off or remove the three **sample_points** layers. Keep the **river, Streets,** and **Homes_For_Sale** layers turned on.

Evaluating Flood Potential

Add Layers

Add to your map the **100_year_flood** shapefile from the **Water_Quality_exercise** folder. Place the **100_year_flood** layer below the other layers. The **100_year_flood** layer is a polygon that shows land area in the region that will be impacted in the event of a major flood, also known as a 100-year flood. Any homes located within the 100-year flood zone will be undesirable properties.

Select Features

With the **100_year_flood** layer active, use the Select Features tool and draw a large circle around the flood zone polygon. When selected properly, it will turn yellow.

Using the Buffer tool, select a buffer distance of 1 foot and choose features from the **Homes_For_Sale** layer. Make the **Homes_For_Sale** layer active and examine the attributes. Properties listed in the Attributes box are located within the 100-year flood zone. Determine which property addresses on Table 1 appear in the Attributes box. For those addresses, write "yes" in the Flood Potential column (E) of Table 1.

Buffer

Attributes

Choose the Properties
You have now investigated the characteristics of properties for sale in the area you wish to purchase a home in terms of river access, land use, water quality, and flood potential. On Table 1, circle the addresses of properties that are suitable for purchase according to the evaluation criteria. To do this, eliminate those properties that have no river access, land use that is something other than residential, cropland, or forest, poor water quality, and vulnerability to major floods.

Table 1: Homes Within 450 Feet of the River

A	B	C	D	E
Address	River Access	Land Use	Water Quality Evaluation	Flood Potential

Table 2: Water Quality Worksheet

A	B	C	D	E
Station ID	Specific Conductance	Dissolved Oxygen	pH	Water Quality Evaluation
Station 1				
Station 2				
Station 3				
Station 4				
Station 5				
Station 6				
Station 7				
Station 8				
Station 9				
Station 10				
Station 11				
Station 12				

What Have You Learned?
In this exercise you have identified housing sites in relation to river access and types of nearby land use. You have also learned about measurable water properties that indicate water quality. You have used GIS tools such as buffering and MapTips to correlate water quality data and flood potential with home locations. Depending on the availability of data, the approach and skills that you have employed in this exercise are ones that might be applicable to a real situation in the future.

Exercise 9: Agriculture and Environmental Quality

Agriculture is one of the largest components of the United States economy. Agriculture activities also have important environmental impacts. Farmers manage soil, water, and biological resources. For many people, farming also provides the foundation of cultural and regional identities. In this exercise we will explore the distribution of major agricultural products and consider some of their potential impacts on groundwater resources—both use and water quality.

The data used in this project are from the USDA-NASS Census of Agriculture, 1997, for the contiguous 48 United States. This census records detailed statistics on numbers of farms, dollar value of production, acres in production, and other variables. Although the data are several years old, the larger production patterns have changed little, so these data are still informative today.

Getting Started

To begin this exercise, copy the **agriculture_census** directory from your CD to the C:\ExploringGIS\folder directory on your computer's hard drive. Start ArcExplorer. Open the project **ag_census.axl** by either using the Open Project button on the toolbar or by going to File > Open Project (Control O) and then navigating to the **agriculture_census** folder.

Open Project

(Remember that if you see an error message while opening the file, just press Continue.)

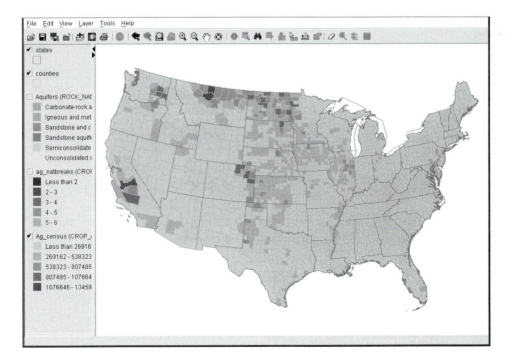

Investigate the Data
When you open this project, it shows the number of acres of agricultural land in each county. To begin, use the Pan and Zoom In tools to investigate the high and low values in the map. List several regions of high values (intensive cultivation).

Answer:_____

_____

_____

What might be produced there?
Answer:_____

Pan

Zoom In

If you can't readily recognize regions or you need help remembering states, use MapTips to show state names: click on the MapTips button in your toolbar. Select the **states** layer on the left side of the MapTips window, and the STATE (name) field on the right, as shown below. Then press the Set MapTips button.

MapTips

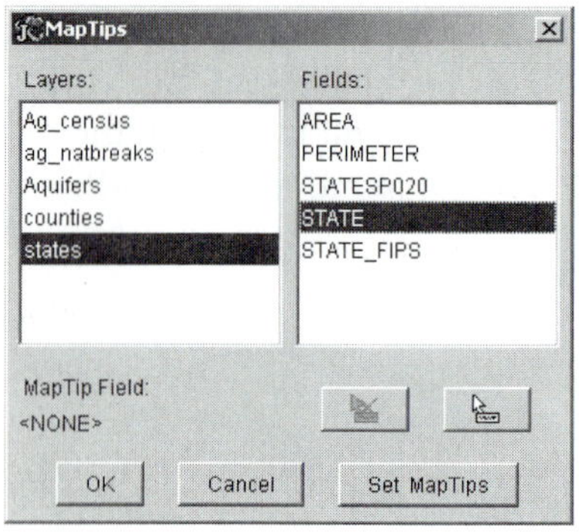

Later, it may be useful to turn on MapTips for other variables, such as corn acreage (CORN_AC) in the **Ag_census** layer. Only one layer can show MapTips at a time, however.

Investigate the Metadata
Data tables often have cryptic column names. What exactly do names such as "Exp000" represent? Often data files are accompanied by documentation, files explaining what the names mean, what units are used, dates data were collected, and other useful information. Files that provide these explanations are often called "metadata," because they are effectively data about data. For example, the metadata might indicate that "Exp000" represents the expenses involved in farm production tabulated by county, in 1997, recorded in thousands of dollars.

Metadata files can be long and extremely detailed. A short version of the metadata, called **Ag_Metadata.txt** is provided. You can find the file in the C:\ExploringGIS\agriculture_census directory, where you found the map files you are using.

Open this metadata file: you can do so using a text editor or a word processing program. Remember that you are opening a file with a .txt extension. Read the explanations of the crop attributes you have been using. How many attributes (variables) are there? What units are used? Keep this metadata file open so that you can refer to it as you work.

Top Producers

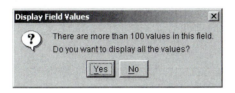

Query
Builder

Let's examine the counties with the most agricultural acreage in the country. First you'll find the top counties, then you'll look at the acreage and products. We'll use the Query Builder to do this. Make sure the **Ag_census** layer is active (click on the layer name in the list of layers on the left side of the map so that a gray box appears around the layer). Then click the Query Builder button in the toolbar. A list of attributes will appear on the left side of the Query window. Click once on CROP_ACRES. A list of values will appear on the right, but first, ArcExplorer wants to know if you really want to display all 4,700 values in the list. You will see this message:

In this case, you DO want to see the full range, because you want to know the top acreage. So click YES.

When the complete list shows up, scroll to the bottom to see the largest number of cultivated acres in one county. What is this largest number? Answer: _____

In the query space (below) click once after CROP_ACRES to move your cursor there. Then click the equal (=) button, and click on the largest crop acreage value. The result should be a line that looks like this:

(If you make an error in your query, you can always click the "Clear" button or overtype parts of the statement.)

Query space →

Click Execute. Then look at the map. You should see the top county highlighted in yellow.

Where is it? Answer:_____

Use the Identify tool to scan its attributes.

Identify

Now modify your query to compare the top TWO counties. In the query space, delete "= 1345807", and replace it with " >= 1250984", the second-largest number. When you Execute the query, you'll again see the resulting counties in yellow. This time, use the Identify tool to compare their values.

	#1 County	**#2 County**
County name, state		
Crop acres		
Harvested acres		
Irrigated acres		
Top agricultural acreage		
Crop value		
Top livestock, number		

How important is irrigation in these two counties?
Answer:_____
_____

Can you explain the difference in crop acres and harvested acres in the Montana county? (Try to imagine what kind of farming is done in western Montana.)
Answer: _____
_____
_____
_____

What is the approximate ratio of total crop values between these two top counties?
Answer:_____

These are only two counties out of over 4,700, but they show some of the contrast in production systems and products among counties.

Find Your Area
Now that you have explored major production regions, take a look at production in your area. If you do not live in the states shown on this map, select a location you are interested in or familiar with for the following section.

Can you find your county on the map? Before looking at the data, do you know, or can you guess, what kinds of crops are produced in your area?
Answer: _____
_____
_____

If you're not certain that you can spot your county on the map, you can use the Find tool. Click on the Find tool, then select the **Ag_census** layer, and type the name of your county. This search is case sensitive, so be sure to capitalize properly. Click the Find button, and a list of options will be returned.

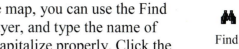

Find

Click on one of the options in the list, and it will be highlighted in yellow on the map. Go down the list until you find your county. Note that the list is arranged generally north to south, so if your county name is a popular one (such as Washington County), and if you are in the South or Southwest, you should start from the bottom of the list.

Once you have found your county, use the Identify tool to learn about agricultural production there. You may be surprised to find some production even in urban counties. Write down the name of your county. What are the major crops or livestock produced in your county? What is the value of crops in your county?

Identify

Answer:_____

_____

_____

Try to identify a county in your state (other than your own) that you think is an important agricultural county. Repeat the Query and Identify process above. Write down the name of your county. What are the major crops listed for that county? (Note that the data do not include *all* crops produced, only a selection of major crops.)

Answer:_____

_____

_____

What is the value of crops in that county, and how does it compare to crop values in your county?

Answer:_____

_____

_____

Investigating Major Crops and Livestock
You have looked at the principal production regions and at your own area. Now you'll examine individual crops.

First, change the **Ag_census** layer to show acres of corn production (CORN_AC). (To open the Properties window shown below, either double click on the layer's name, or click the Layer Properties tool in the toolbar.) Make sure that the "Remove Outline" box is checked. Outlines obscure the pattern in the map, since many counties are minute at this scale. Colors should be scaled from a very light blue to dark blue. If necessary, you can adjust the "start" and "end" colors to ensure good visibility as you work on this section.

Layer
Properties

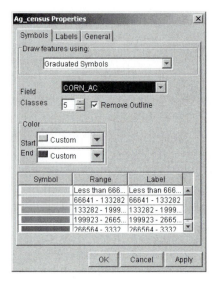

To see the patterns easily, turn off the gray **counties** layer, listed near the top of the layers list.

Which five or six states have the most acres in corn production?
Answer:_____
_____

Now change the display to show acreage in soybeans (SOY_AC) using the same procedure as you used for corn acreage. Which states are leaders in both corn and soy acreage? Do any additional states show up among the top soy producers?
Answer:_____
_____
_____

Corn and soy are among our top agricultural commodities, in volume and total value. Soy and corn are also notable because they are leaders in genetically modified crop production. The principal reason for modification of these crops is for herbicide tolerance or pest control (for example, "Roundup ready" soybeans are modified to tolerate the herbicide Roundup, which allows spraying for weeds without killing soy crops; "Bt corn," containing genes for *Bacillus thuringiensis,* produces chemical compounds that kill insect pests).

Let's look at soy production in more detail, since this is one of our top agricultural commodities. Make sure the **Ag_census** layer is active, then select the Query Builder, which you used above. Single click on SOY_AC. When asked if you'd like to display all the values, click Yes. Then scroll down the list and see the largest value in the table.

Query Builder

What is the greatest number of acres of soybeans listed?
Answer:_____

Let's see where the big producers are by running a query for counties with more than 100,000 acres in soybeans.

Your query should look like this: (SOY_AC > 100000)

This query displays the top soy producers a little more clearly than before.

Now do a query to see which counties produce more than 10,000 hogs.

Your query should look like this: (HOGS > 10000)

Do you see a similar pattern to soy production? Which states are the primary hog producers?
Answer:_____

If you eat bacon, ham, sausage, or other pork products, which of these states is the closest source to your grocery store? How might you find out the source of the meat products you eat?
Answer:_____

_____

_____

Now look at counties with more than 1 million chickens: CHICKENS >= 1000000
The distribution of hogs, chickens, corn, and soybeans are notable because corn and soy are used more for livestock feed than for human food. Thus, it is often efficient to produce livestock near the crop fields. Did you observe these patterns with chickens and hogs? _____

Compare the Distributions of Major Crops
In this section, you will display the agriculture data using a variety of the attributes (fields) in the table.

Queries are a good way to investigate the distribution of agricultural production, but they can be slow. To speed up the comparison of variables, we have provided a layer that groups crop production by natural breaks in the data set (**ag_natbreaks**). The file you have used thus far breaks data into classes by equal intervals; that is, each class has the same range of data values (see drawing below). Since natural breaks distribute the data values more evenly among the colors, it provides a clearer view of the geography of production. Use the **ag_natbreaks** layer in the sections that follow.

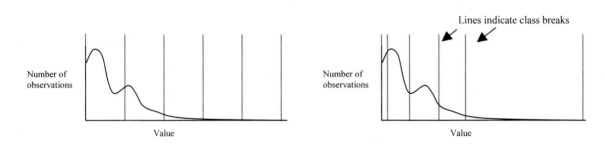

Equal interval and natural breaks classification methods. The vertical, gray lines on both graphs indicate class breaks. In this data set, equal interval assigns just two colors to nearly all observations, leaving three colors largely unused. Natural breaks (at right) uses all five colors more effectively.

As you work, you can examine different attributes quickly by clicking the Apply button in the Properties dialog box, rather than OK. The Apply button applies changes without closing the window.

To begin, quickly repeat the display of corn and soybeans using the **ag_natbreaks** layer. Also inspect wheat and cotton. Identify six leading states in the production to fill in the following table. (States low on the list may be hard to distinguish: just choose one.)

Corn	Soy	Wheat	Cotton

Can you explain why these crops are grown where they are? What do you think might be the primary environmental requirements of these crops? If possible, discuss this with other students before answering.

Answer:_____

_____

_____

List some of the uses of these crops. Which make up the bulk of American agricultural production?

Answer:_____

_____

_____

Is there any correspondence between your previous livestock findings and crops above?

Answer:_____

_____

Now look at the distribution of irrigation. List several states or regions that have the most irrigated acres.

Answer:_____

_____

Are any of these states among the primary producers of corn, soy, cotton, or wheat?

Answer:_____

Print and Explain a Map

Choose one commodity from those you have examined above. Make a copy of the map using the Copy Map To Image tool. Open a word processing program such as MS Word, and paste the image into a new document.

Copy Map
To Image

Below the image, type a paragraph explaining environmental factors that you think help explain the distribution you see in your map. If possible, discuss potential explanations with other students, but use your own words to explain the map. Print your document, and hand it in with your exercise.

Examine Groundwater and Agriculture

Turn on the **Aquifers** layer. This file shows major near-surface rock formations that provide well water. In general, aquifers are cracked or porous rock formations; water collects in and moves through the cracks or pores. Water can accumulate or travel quickly in rocks that are deeply broken and cracked or in unconsolidated material, such as sand or gravel. Wells in these formations can be susceptible to contamination from agricultural chemicals or nitrogen from livestock manure. Here is a brief description of the important characteristics of these formations for our purposes:

Carbonate rock aquifers: composed mainly of limestone. In some cases, acidic runoff dissolves the stone, expanding cracks and making these rocks extremely porous. As a consequence, water and contaminants may flow quickly through the ground.

Igneous and metamorphic rock aquifers: these rocks may have many cracks, although they are often less porous and certainly less soluble than most other rocks.

Sandstone aquifers: made of sandstone, in which cracks are sometimes substantially eroded and enlarged by moving water.

Unconsolidated and semiconsolidated sand and gravel aquifers: often composed of relatively loose material that allows water to travel quickly. Like limestone, these aquifers may be susceptible to groundwater contamination from agricultural chemicals and from manure.

Examine the map. To see the names of major aquifers, turn on MapTips for the **Aquifers** layer (as shown below).

MapTips

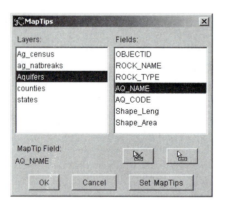

Roll your mouse over the map of aquifers. Identify some of the larger aquifers.
Answer:_____

Does the map show a major near-surface aquifer near where you live? Does it have a name or a description?
Answer:_____

_____

Now use a Light screen fill (as shown below) to display rock types on the **Aquifers** layer. This semitransparent screen allows some visibility of the agriculture layer below. Colored boundaries around major near-surface aquifers identify which type of rocks make up these water-bearing layers.

Set the **ag_natbreaks** layer to show irrigated acres. Using the MapTips for aquifer names as you did above, can you see a correspondence between irrigation and any of the aquifer formations?
Answer:_____

MapTips

It might be easier to see the relationship between irrigation and aquifer distribution if you do a query on the actual values of irrigated acres.

Query Builder

Make the **Ag_census** layer active once again, and find counties with more than 100,000 acres in irrigation. Your query should look like this:

After executing this query, examine the map. It should now be easier to identify which aquifers underlie heavily irrigated areas. (Note that some of these aquifers, such as that in California's Central Valley, may have been used historically and may be largely depleted today.) List the primary aquifers that coincide with heavily irrigated farming regions. Does your list look the same with this different display method?

Answer:_____

_____

_____

Of course, not all irrigated crops use groundwater. Often rivers are more accessible water sources. The basaltic aquifers of the Pacific Northwest, for example, are generally less important than surface waters for agriculture.

Aquifers are also important sources of drinking water. But intensive agriculture can introduce farm pesticides and herbicides to groundwater, especially in unconsolidated, semiconsolidated, and carbonate (limestone) aquifers. Focus for now on the High Plains (Ogallala) aquifer in the center of the United States. This aquifer is a complex of semiconsolidated and unconsolidated regional rock and sand formations extending from Nebraska to Texas. Since it is relatively shallow and wells are relatively easy to drill, the High Plains aquifer has been largely depleted in many areas, and it has been the subject of ongoing policy debates about groundwater policy. Let's examine agricultural production over and around the High Plains aquifer.

Use the Query Builder again to search for counties with more than 50,000 acres of soy, corn, cotton, and wheat. Although 50,000 is an arbitrary number it does identify the peak distribution of these crops.

Query Builder

Your queries should have this form: (COTTON_AC >= 50000)

Do any of these crops correspond to the distribution of the High Plains aquifer?

Answer: _____

Now run queries to examine the distribution of hogs, cattle, and chickens. Again, use an arbitrary value of 50,000. Hogs and cattle, in particular, produce large amounts of waste when kept in feedlots or livestock barns.

Your queries should have this form: (HOGS >= 50000)

Which of these types of livestock needs to be watched most carefully to prevent contamination of the High Plains aquifer?

Answer:_____

Can you identify other groundwater systems, elsewhere on the map, that could be threatened by hogs if manure is not carefully managed?

Answer:_____

_____

What Have You Learned?
Agricultural production is central to the American economy. As you have observed, agriculture is widespread and diverse. The farming economy also has important environmental impacts. One of these is on groundwater consumption and groundwater quality. Additional environmental considerations include surface water, soil erosion, habitat distribution, and biodiversity impacts. These issues can also be addressed with spatial data. Although we have fewer legal controls on farms' effects on the environment than we have on factories' effects, there are many farmers who have invested considerable effort in reducing the environmental costs of production.

In this exercise, you have practiced a number of important skills. You have used the Query Builder to investigate the distribution of values in the data table that is linked to the map. You have modified symbols, making transparent fills and turning outlines on and off, in order to control visibility of patterns on the map. You have used MapTips to investigate the names of features. You have also used the Find and Identify tools to investigate the attributes of counties on the map. In addition, you have copied a map image to a word processing document and proposed environmental explanations of the patterns on the map. All these tasks are essential GIS functions having to do with exploring data, learning about production systems that affect your life, and communicating your findings.